W9-CHH-127

# ACID RAIN
## The North American Forecast

# ACID RAIN

## THE NORTH AMERICAN FORECAST

# Ross Howard
## and Michael Perley

 **Anansi**

Copyright © Ross Howard and Michael Perley, 1980

All rights reserved. Except for brief quotation for purposes of review, no part of this publication may be reproduced, stored in a retrieval system, or transmitted in any form or by any means, electronic, mechanical, photocopying, recording, or otherwise, without the prior written permission of the publisher.

Cover design and illustration layout: Peggy Heath
Typesetting: Imprint Typesetting
Printed by: The Hunter Rose Company

Published with the assistance of the Canada Council and the Ontario Arts Council, and made in Canada by
                    House of Anansi Press Limited
                    35 Britain Street
                    Toronto, Ontario    M5A 1R7

1 2 3 4 5 6        86 85 84 83 82 81 80

*Canadian Cataloguing in Publication Data*

Howard, Ross, 1946–
    Acid rain

Bibliography: p.
Includes index.

ISBN 0-88784-082-5 bd.

1. Acid rain — Canada.    2. Acid rain — United States.
I. Perley, Michael, 1946–      II. Title.

TD885.H68        363.7'392'097        C80-094770-3

# Contents

*There ascended a steady pillar of smoke ... a poisonous pungent sulphur smoke, poisoning the air wide around so that no one will get there without hardship. This corrodes the earth so that no herbs can grow around it. The ground was completely naked one quarter mile away, consisting of large loose rocks, as if thrown on sterile gravel ground, ...*
—Carl von Linne, *Iter Delekarlicum, 1734*

*The answer is blowin' in the wind*

—Bob Dylan

# Foreword

When the idea of this book first came up a year ago, statements by politicians and some scientists gave the impression that acid rain was a brand new phenomenon, sketchily understood. In researching this book, it soon became clear that a broad trail of reports, studies and warnings stretched back almost a decade. I've tried to keep the scientific jargon and complexities to a minimum, although some numbers, costs and specific measurements are necessary to define clearly the realities of acid rain. While describing acid rain and its effects, I've also tried to report on those who were responsible for North America's exceptionally slow awakening to a problem now declared a national disaster in Canada and an environmental nightmare everywhere. The economics of controlling acid rain are not as complicated and intimidating as many people claim; the costs of doing nothing are already more obvious and far greater than most North Americans realize. Enough is already known to begin applying solutions to the acid rain problem but the political decisions to do so remain neglected. Ultimately the solution is up to all of us.

It would not have been possible to write this book so soon without the resources provided by my assignment to environmental affairs for *The Toronto Star*, and the time granted by Ray Timson to get the manuscript underway. Thanks too are due the civil servants and scientists who cooperated, and the friends who supported. They made both jobs easier. The contribution of the Ontario Arts Council is gratefully acknowledged. Jim Polk of House of Anansi Press proved invaluable. Without Mike Perley, this book could never have been written. Without the tolerance and support of Peggy it could not have been finished.

RKH

I wish to thank the Canadian Environmental Law Research Foundation for providing me with considerable material and moral support. My thanks also to the many civil servants and others engaged in the acid rain debate who gave freely of their time and information, sometimes when it was not in their interest to do so.

To Ross Howard, whose unfailing good humour and patience were essential to getting us through the masses of often contradictory and confusing information which has been generated about acid rain.

Finally to Lisette, William and Benjamin, for their encouragement and welcome interruptions.

MP

# Preface
# A Hard Rain

Nellie Lake is a peaceful, sparkling lake in north-central Ontario, part of the protected wilderness of forest and pink granite bedrock known as Killarney Park. The lake and landscape were first captured on the canvases of the Group of Seven's A.Y. Jackson in the 1930s and are now reproduced in prints and calendars for banks, offices and living-room walls across the country. Nellie Lake is typical of thousands of other lakes in Ontario which draw more than one million cottagers, tourists, fishermen and outdoor enthusiasts north each year, away from the urban-industrial sprawl to an unspoiled playground and retreat which stretches across Ontario from the shore of Lake Huron to the Quebec border. It's the kind of lake chosen by the Canadian government for tourist ads in US magazines, a kind of landscape identified worldwide as uniquely northern and Canadian.

Nellie Lake is also "acid dead." The water is sparkling clear because acid has destroyed everything in it, including the color. It is abnormally peaceful because its natural aquatic life, from fish to crayfish to snails, has ceased. Even the frogs and salamanders which once inhabited the shoreline are gone, and fewer insects dart the surface or teem in the depths below. Nellie Lake has been dead less than 20 years; it did not die a natural death. Yet it lies dozens of miles from the nearest permanent settlement, in the middle of a protected forest. No big city stands nearby to dump contaminants into the lake and turn it into a sewage lagoon like Lake Erie or the Hudson

River. There are no roads near the lake and there have been no chemical spills or abandoned mountains of leaky mine tailings. The acid which killed Nellie Lake came from the sky, in a corrosive, poisonous rain of sulphuric and nitric acid.

Within 50 miles east and north of the Killarneys there are at least 140 lakes now known to be dead from acid rain. One hundred and twenty miles to the southeast at Dorset, on the boundary of Ontario's famed Algonquin Park and amid lakes which have drawn city-dwellers to chalets and summer mansions for more than half a century, researchers are charting the day-by-day death of another lake. Back to the west again on the shores of Georgian Bay delicate instruments are housed in white boxes with louvred sides, like miniatures of the thousands of cottages which dot the shoreline. The instruments measure water in the streams and rivers flowing into the Bay and find it to be seriously acidic at certain times. In that swath of customised wilderness which stretches across north-central Ontario—the cottagers call it "up north, Muskoka, the Bay, The Park"—there are more than 48,000 lakes already suffering the impacts of acid rain. Further north, beyond Temagami and Sudbury where the real north begins and reaches towards Hudson Bay and Manitoba, there are hundreds of thousands more. Acid rain falls on them too.

Ontario is not alone. In the Adirondack Mountains of New York state, half of all the lakes above the 2,000 foot elevation are acid dead. On Nova Scotia's eastern-most shores, almost every river flowing to the Atlantic Ocean is poisoned with acid. Further south in Maine, New Hampshire and Vermont the acid rain is falling. And in Florida too. In Minnesota and further west to the Continental Divide of the Colorado Rockies, acid rain is increasingly common. In fact, a steady shower of weak but destructive acid is falling on *all* of eastern North America now, from Hudson Bay to the Gulf of Mexico.

Acid rain may be the most painful example of the old truism: what goes up must come down. Acid rain is the step-child of pollution: more than 30 million tons of sulphur dioxide and nitrogen oxides now spewed annually into the skies over eastern North America's industrial heartland, from the fossil-fueled power plants, the smelters, and the automobile engines. The oxides mix with water vapor in the air and gradually turn to sulphuric and nitric acid. Blown by the winds, much of the acid comes down hundreds and even thousands of miles from its smoke-stack sources, particularly to the north and east, in Ontario, Quebec, the Adirondacks and the Atlantic coast. In southern Ontario, about 11 ounces of sulphuric acid rains on every acre of land, every year.

It has taken only 30 years of this chemical deluge to jeopardize a basic process of nature which has functioned for eons. When the rain falls now, it trickles through soils which have already been exhausted of their own neutralizing chemicals by the three-decade buffeting of acid. Particularly in the northeast, where the soils are thin, the lakelands now shed their acid burden directly down to the rivers and lakes. And the process of acidification accelerates; as little as five more years of acid rain could finish off the last of the neutralizing chemicals in some leached soils and waters. The lakes and reservoirs which eventually collect the flow are evolving into chemical sinks and sewers. In the near-north recreation wilderness of Ontario, thousands more of the 48,000 lakes may be acid dead within 10 years.

Acid rain is killing more than lakes. It can scar the leaves of hardwood forests, wither ferns and lichens, accelerate the death of coniferous needles, sterilize seeds, and weaken the forests to disease, infestation and decay. Below the surface, the acid neutralizes chemicals vital for growth, strips others from the soil and carries them to the lakes and literally retards the respiration of the soil. The rate of forest growth in the White Mountains of New Hampshire has declined 18 percent between 1956 and 1965, a time of increasingly intense acidic rainfall.

Acid rain no longer falls exclusively on the lakes, forests, and thin soils of the northeast; it now covers nearly half the continent. There is evidence that the rain is destroying the productivity of the once-rich soils themselves, like an overdose of chemical fertilizer or a gigantic drenching with vinegar. The damages of such overdosing may not be repairable or reversible. On some croplands, tomatoes grow to only half their full weight, and the leaves of radishes wither. In Ontario, tobacco leaves are pitted and made unsalable by the action of acid rain and associated airborne pollutants. And naturally, it rains on cities too, blistering automotive and building paint, eating away stone monuments and concrete structures, and corroding the pipes which sluice the water away to the lakes and rivers again. In some communities, the drinking water is laced with toxic metals freed from the metal pipes by the acidity. As if urban skies were not already grey enough, typical visibility has declined from 10 miles to 4 miles, along the eastern seaboard, as acid rain synthesizes into new smogs. And now there are indications that the components of acid rain are a health risk, linked to human respiratory disease.

Little of this should come as a surprise to North Americans, particularly Canadians, and the scientists, bureaucrats, and politicians responsible for protection of the environment. The first warning came in the early '70s from Sweden, the result of 20 years research into air pollution blown to Scandinavia from the factories and cities of Europe and England. The Swedish report cited massive fish kills and a death toll of lakes numbering in the hundreds. It pointed to crop and forest damage, property corrosion, and health implications, concluding that acid rain meant an environmental disaster for Sweden in the near future. And it added: "a similar situation might possibly exist within certain regions of Canada and the north eastern parts of the USA. ... A detailed study of the likelihood of such a development is a matter of urgency." The report was Sweden's key contribution to the 1972 United Nations Conference on the Human Environment, the Stockholm conference heralded as the awakening of a new ecological consciousness. The report was printed in English.

On this side of the Atlantic there was little attention paid to the Swedish alarm. Some scientists knew of the Scandinavian studies, others should have known them as part of their routine responsibilities for environmental protection. Collectively most scientists failed to tell anyone else—particularly the public. Until the mid 1970s the sum of North American research into acid rain's widespread implications amounted to a drop in the bucket. And that single drop was 90 parts Swedish research and 10 parts concern that the same conditions seemed to prevail in North America. In Canada at least one academic scientist studied the situation and sounded his own warning, but he was largely rebuffed and ignored. Today, as a leading researcher into acid rain, Harold Harvey says "it's almost as if government scientists and decision-makers didn't want to have to start facing reality." The myth of a clean, green northland dies hard, and three Canadian mass circulation publications rejected Harvey's findings. One publication denied their veracity because a staffer once "caught a fish in that area." That area was Killarney Park, and some of the lakes there had been dead for a decade.

It wasn't until 1978 that an Ontario government study got around to confirming the Swedish example. A quick survey of 200 lakes in and near the Killarney region showed 40 were dead and 100 more were on the critical list. That news made the deep-inside pages of a few newspapers. But one year later the death toll of lakes was translated into a death knell for much of Ontario's cottage country. In 18 sessions of public hearings by a committee of the Ontario Legislature, the range and ramifications of acid rain were finally sketched out by cautious bureaucrats and highly critical academics and environmentalists: the extinction of fish species, the weakening of soil, the threat to vegetation. It was noted that the Adirondacks were similar in geography, acid rain, and damage reports. There were even references to the Swedish studies, although too often without crediting them for the warning to North America they contained. And towards the end of the hearings a Liberal (opposition) party member raised the question: "Isn't it true we're facing the environmental disaster of the decade?" No-one denied it. South of the border the proceedings of the Ontario committee went almost unnoticed, as was acid rain.

After those committee hearings concluded, acid rain continued to fall and the damages continued to mount. And they will increase yet again. To June 1, 1980, nothing had been accomplished in Canada or the United States which signalled any measurable reduction in acid rain for the coming decade. In that decade, the lakes-death toll across North America will rise substantially, and the wider-ranging damage may hit hard at forestry and agriculture. Yet both Canada and the United States are now busily engaged in the construction of additional sources of acid rain pollutants.

In Canada, to 1980, the largest sources of acid rain, which are in Ontario, remained under no constraints to reduce their emissions. The greatest source of them all, the world's-largest International Nickel Company (Inco Ltd.) smelter in Sudbury, upwind of Killarney Park, was freed from previous pollution cut-back orders in 1978 and allowed unchanged emissions to 1983. The other major offender, the provincial Ontario Hydro utility, remained opposed to installing pollution controls and continued building another uncontrolled new source. Early in 1980 minor amendments were made to pollution control orders on Inco but substantial abatement was put off for some future time. South of the border, more than 200 of the dirtiest fossil-fueled power plants on the continent remain free to spew their oxides until the day the plants die, in 20 years or more. Those often privately-owned plants lie clustered close to the source of their fuel, which is eastern coal. Behind that coal are powerful political and economic interests which have already defeated previous attempts at regulation and show no signs of buckling under to more. Those plants form the rim of a funnel which blows air to Ontario, the Adirondacks and the eastern seaboard bearing some 20 million tons of acid rain-causing material—more than double the volume which blighted Sweden, six times the Ontario total. But there's more to come. In its rush to a lessened energy vulnerability, the US is going to mine more coal for new power plants, and for synthesized fuel, over a widening area. The $142 billion energy scheme of President Jimmy Carter in July 1979 means more acid rain. The new power plants will have new controls, tougher than existing plants, but the eventual total output is still going to be an increase in sulphur and nitrogen oxides.

Since early 1978 there have been desultory expressions of government concern about acid rain on both sides of the border, although the irony of the first American complaint is not lost on Canadian diplomats. That complaint was based on a US Senate resolution in May 1978 urging Canada to curb a minor emission of acid rain from two utility plants in Ontario and Saskatchewan. Even then it was estimated that US total sources outnumbered Canadian five to one, and up to half of the total American pollution was wind-blown north across the border. However, two years of transboundary meetings and sometimes acerbic scientific exchanges have produced nothing beyond an agreement that an international agreement is needed to curb acid rain sources in both countries. Scientific data has proven the obvious: the rain blows both ways. The acid rain damage is mounting. And neither country will go first in shouldering the costs of a clean-up.

Despite political and bureaucratic promises that Canada, and particularly Ontario, must and will move against local sources of acid rain, and that the US will be "encouraged" to do the same, the substance of Canada's position to 1980 remained a deceptive assurance and desperate hope that the US would move first. To support their inaction, Canadian leaders—and scientists and industrialists—argue that a Canadian clean-up alone is valueless since the country emits fewer total oxides than the US and suffers great imports of American fumes. And south of the border, the fingers point north. New Canadian abatement policies are needed, before any new legislation to put the lid on American pollution sources can be pushed through an energy-preoccupied Congress. Canadian protest to date has been inconsequential. The problem of acid rain isn't known, to most US legislators. The single small step Canada took in early 1980 to curb one source signalled no change in American awareness.

Ultimately it will matter little which country goes first. Both must act. It may cost as much as $10 billion in Canada and 8 times as much in the eastern US to curb—not completely eradicate—acid rain over a 20 year period. The technology is largely available now, although in both Canada and the US it has remained a victim of traditional corporate and political blackmail threats that a clean-up will cost jobs. However, not cleaning up will cost jobs too. To 1980 Canada had not bothered to research the potential economic losses from acid rain in any detail. Ontario, the province ostensibly hardest hit, had not spent a cent to calculate the jobs to be lost, for example in a devastated $120 million per-year fishing industry or $900 million-per-year tourism industry in acid-impacted areas, or a $2 billion forest industry which could be facing a 15 per cent decline in productivity. Unable to quantify losses to the final decimal point yet, politicians and bureaucrats have preferred to say nothing and issue no warnings. But simple calculations of the value of what is at stake, coupled with already-known and increasingly obvious measurements of the sensitivity of such resources, are alone more than enough proof that acid rain is an unacceptable economic risk for Canada. And beyond the financial losses are the less tangible but critically important clean lakes, healthy forests and fresh air which mean so much to Canadians and Americans.

Both Canada and the US have allowed claims of clean-up costs to mount higher, unchallenged, while neglecting both the physical and economic realities of acid rain damage. In Sweden they manage things better. The government there in 1972 pointed out that an unchanged fall of acid rain could possibly cost a 7 per cent loss in forest productivity, a 50 per cent loss of living lakes, a $45 million loss from property corrosion, and a increased human mortality by the year 2,000, and judged the risk unacceptable. Sweden introduced laws in 1973 to reduce its acid rain sources, and began pressuring Europe to curb its emissions.

But in North America, with a far more costly pollution problem, Canada and the US do little but yawn, and wait. Canada faces no alternative but to act first, in substantial ways, given the traditional impossibility of subtly influencing the course of its lumbering and largely self-interested neighboring giant. The giant so far has been uninterested in acid rain but the cost and destruction is mounting quickly and cannot be long ignored. Unfortunately, to wait any longer at all may be too late to avert the worst of the damages and costs. Armed with enough information now, citizens in both countries can choose what to do about acid rain. The choice is ours, because we are responsible for it, and it won't go away by itself.

# 1
# The Dying Lakes

In the spring of 1966 a University of Toronto zoologist named Harold Harvey supervised the introduction of 4,000 young pink salmon into Lumsden Lake, a small body of water in the centre of the Killarney Park wilderness. Harvey, a fisheries specialist, had read of the survival of pink salmon in Lake Superior and was interested in the possibility of breeding these fast-growing, popular sports fish in lakes elsewhere. The 45-acre lake he chose was surrounded by the low, quartz-streaked hills of the La Cloche Mountains, just inland from Georgian Bay and roughly 200 miles northwest of the university, It was an appropriate lake, cut off from any road, with no-one living near the lake or its five small upstream headwater lakes, but still a fisherman's dream for lake trout, perch and lake herring, according to government surveys in the early 1960s. Water flowed into the lake over a 10-foot waterfall from upstream, and the only exit was through a sieve of logs and debris and over a 45-foot waterfall. The clear, 60-foot-deep waters were perfect for the salmon.

As an extra precaution Harvey strung wire screens across the outlet, to keep his experimental fish population in place. While checking his handiwork, Harvey netted some white suckers, a common fish in the area and potential food for his salmon. The suckers were abnormally small, and seemed to be unsuccessful at breeding. He hauled them back to his laboratory for later study of some possible genetic or reproductive failure. Over the winter Harvey queried provincial government fisheries officers across the province for any more news about his discovery, and received a half-dozen vague and unconcerned confirmations. He set the replies aside by the spring of 1967 when it was time to check on his salmon population. He spent much of that summer looking for them, and found not a single one. All 4,000 fish had disappeared. In dragging huge nets through the lake, he found only more dwarf suckers, and only old ones at that. There were no young. And yet, as tests showed, the lake had enough food for many more fish. The loss of all the salmon was unexplainable, as was the aberrant population of dwarf suckers.

In the spring of 1968 Harvey and his students returned to Lumsden Lake, and netted and tagged more than 100 suckers there and 60 more in a neighboring lake, selecting the fish to represent all ages and classifications. And by the following spring, only a handful of the tagged suckers could be recaptured. In Lumsden Lake there were no one-year-old suckers at all, and females were not reproducing. Harvey's team promptly netted and tagged an estimated 25 per cent of the entire sucker population in the lake, and set them free in the lake again. By the fall of that year, endless net-dragging produced only one tagged fish. Worse, none of the lake trout and perch and herring which local fishermen had boasted about only five years ago turned up in the nets. From eight reported species of fish, the lake had been reduced to one—suckers.

During the summers of 1969 to 1971, Harvey and his researchers lugged their canoes upstream to the five tributary lakes, and overland to some neighboring ones. The 1,500 pounds of nets, tags, water sample bottles and other equipment also had to be carried—no float-equipped planes were allowed into Killarney Park. And the back-breaking work produced heartbreaking results: the upstream lakes were almost devoid of fish, as was nearby George lake. Also barren was OSA Lake, named after the Ontario School of Art which had fostered so many members of the Group of Seven between painting trips to Killarney four decades earlier. Once planning the introduction of a new fish population in one small lake, Harold Harvey was now charting the disappearance of virtually all fish in several lakes, lakes which only a decade earlier had been thriving. What Harold Harvey had discovered was the decline and death of the Killarney lakes by acid rain. But it was only the beginning.

Along with all the other paraphernalia Harvey and his crew portaged overland to the lakes was equipment to measure acid levels in the water. It was a standard routine to test acidity, and government tests a decade earlier were available for comparison. Lumsden Lake, for example, had shown an acid strength of 6.8 on the pH scale, back in 1961. The pH scale is a simple gauge running from 0 to 14. Perfectly neutral water is pH 7 (as in distilled water in a laboratory); higher numbers are called alkaline or basic—baking soda is pH 8.5 and ammonia is pH 12. The lower numbers are acidic—vinegar is pH 2.5, limes are 1.7, and battery acid is well below pH 1. The pH scale is deceptively simple: it works on a logarithmic progression: a change by 1 unit is a 10-fold change. A pH 6 is 10 times more acidic than neutral pH 7. In other words, the lower the number on the scale, the very much greater the acidity. A minor deviation from neutrality is normal; even natural rainfall in most of the world is slightly acidic due to traditional chemicals in the air, and natural rain has a pH of 5.6, almost everywhere in the world. By the time this slightly acidic natural rain has washed across the land, other chemicals have counter-acted and neutralized it, and the water ends up much closer to neutrality—pH 7—or even slightly alkaline when it finally collects in lakes and rivers. It is a process which has been going on since virgin rain first fell on earth. But in the Killarney lakes region in the 50's the natural process began to change dramatically.

As Harvey's acid measurements revealed, the pH of Lumsden Lake had fallen from 6.8 in 1961 to 4.4 by August 1971—a highly abnormal 100-fold increase in acidity in one decade. The upstream lakes and the nearby ones which had been tested mainly for fish and coincidentally for acidity had also suffered major declines in pH (thus major increases in acidity), and major losses of fish. By 1972 Harvey had charted more than 60 lakes in the Killarney region watershed. And lake after lake had a low pH and low fish population. Different species were wiped out in different lakes, some lakes held big fish and no young, and some were simply empty. According to records, almost all of these lakes had been healthy a decade earlier. Harvey and his chief research assistant returned to the university and began preparing a report on their findings.

What Harold Harvey set out to describe concerning Lumsden Lake is basically the way fish and lakes die from acidification anywhere. The process and pattern is not exactly the same in every case, but the end result is: loss of some fish, loss of a fish species, and eventually the loss of all fish. The invisible changes begin when the acid's hydrogen ions begin disrupting the chemical bonds among calcium, magnesium and other elements in water. This is the stuff of high school chemistry. In simple terms, acid produces chemical changes in the blood of the fish, and their basic body metabolism is altered. The strangely dwarfed white suckers Harvey discovered in 1966 were suffering metabolic disruption, and their failure to breed was the result. Starved of calcium, female fish were unable to produce and release the ova for fertilization by the males. Nor, in some cases, did the fish reach proper maturity—they hadn't grown at a normal rate because their respiratory systems were unable to process oxygen in water passing over the gills, and the blood was unable to transfer the protein to the flesh and bones.

But some of the white suckers were more than just dwarfed, poor reproducers; they were also deformed. By 1972 nearly one-third of the netted and examined fish showed twisted and arched backbones, flattened heads and strangely curved tails. Their bones were undersupplied with minerals essential for spinal strength. In other cases, their bones were slowly dissolving, as the blood drew the calcium from the bones to replace what no longer was available in food in the acid waters.

Admittedly, not all the fish were suffering. With fewer and weaker fish competing for food in Lumsden Lake, some white suckers prospered and grew bigger. But they were growing older too, and with few if any young fish to replace them, they were in danger of growing extinct. However, white suckers (and the 4,000 salmon) were not the only fish which once thrived in Lumsden Lake. In the 1950's that lake had held at least eight species of fish. By 1960 yellow perch were no longer around, nor burbot. The last lake trout was caught in 1967, the last trout, perch, and herring 1969—the same year white suckers were confirmed to be on the wane—and by 1971 with intensive net dragging only occasional lake chub could be captured. But nothing else. As is well documented by research now, acid is selective in its decimation. Not all fish are equally vulnerable, at least initially. Fish decline in a particular order as acid intensity increases.

At the top of the vulnerability list come the fishermen's prized brook and rainbow trout. When the pH falls below 6.5 they begin to suffer: the females still lay eggs but the eggs don't hatch, poisoned by the acid in the water. The surviving but older population heads for extinction. Below 6 it is downhill all the way. Some fish fail to reproduce, others produce eggs which fail to hatch. By pH 5 smallmouth bass, walleye, and lake trout populations have no future. And finally, somewhere around pH 4—acid 1000 times stronger than neutral water—there's little left in the lakes but the garbage fish, the chub, rock bass, pumpkinseed and lake herring.

By the end of 1972 Harvey's research team had grown, and the total of surveyed lakes had risen to 150 in the 800 square mile Killarney region. Almost half the lakes had acid levels below pH 5.5, including 33 which were "critically acidified" below 4.5. Some, including one ironically named Acid Lake, held no fish at all. It was clear that the isolated, untravelled lakes had suffered no direct pollution or dumping of acid, and the researchers had turned their suspicions to the air. It was already known that a giant nickel smelter at Sudbury, 50 miles to the northeast, had seriously poisoned its adjacent lakes years ago with air pollution fall-out. And it was known that some of that fall-out included sulphur dioxide contaminants which turned to sulphuric acid in those Sudbury lakes. So the results of individual samples of falling rain taken during Harvey's 1969-1971 Killarney research were examined again, and confirmed the suspicion: rainfall over the Killarneys was itself acidified, and coincidentally, at least some of the rain had blown from the smelter area according to regional weather maps in Sudbury. Most of the rain readings were in the pH 4 range, an acid level which could be expected to have gradually contributed to declining pH in the lakes as the rainwater flowed into them, but hardly so soon as to have wiped out an entire year of young fish all at once, as was seen in Lumsden and other lakes. Snow samples revealed the answer. Tests for acidity taken in a stand of birch trees alongside Lumsden Lake, in mid-February 1970, had revealed the top 10 inches of snow were acidified to pH 3.3. Deeper, the acidity was even greater.

The same kind of conditions existed around other nearby lakes. And mid-winter tests the following year had shown acid snow. As Harvey speculated in his report "Acidification of the La Cloche Mountain Lakes", when that snow melted, as it does quickly when spring and the ice break-up hits northern Ontario, acid deluged the lakes and rivers in a sudden shock.[1] For a matter of days, the rivers ran with more acid than any typical or average rainfall could cause. And spring thaw is the key breeding and spawning time for many fish. Acid shock loading, as it is called now, is the cruellest stage of acidification—it hits the lakes when their aquatic dwellers are most vulnerable, and with deadly results. With the spring floods over, a more normal acidification resumes, and the surviving adult fish grow old, and the species slowly declines to extinction. An acid sampling of lake water in mid summer may show a serious but still tolerable pH 5 or greater but the damage has already been done.

When Harvey published his first report, he concentrated on the death toll he'd observed in the Killarney lakes, and its immediate causes. But he also referred to Swedish studies of acidification from industrial sources, and studies of the impacts of the Sudbury smelter confirmed in Sudbury-area lakes in 1960. In those days it was generally believed that acid contaminants from sources such as smelters survived in air only for 10 hours, and the Sudbury study had examined lakes only within a 15 mile (or 10 hour) travel time for air-borne pollution. The Swedish studies, Harvey noted, showed pollution could last longer and travel further. The Killarney lakes were further downwind from Sudbury, he added. "It is possible for large amounts of sulphur dioxide released in Sudbury to be carried into the La Cloche Mountains area." Sulphur dioxide turns to sulphuric acid in water, and the smelter had been putting 2.6 million tons of sulphur dioxide into the air per year, in the period before Harvey wrote his report. He added that soils of the La Cloche Mountains were unusually low in acid-neutralizing minerals, which could mean an even stronger acid would run into the lakes when the soils became exhausted. There were many factors worth urgent study in the acid-impacted Killarney Lakes region, Harvey concluded. The lakes were dying.

As later events and belated research has proved, Harold Harvey was completely correct in his conclusions, with one exception. Those factors of acid shock, airborne pollution and weak soils apply not only to the Killarney area but also to lakes across much of central Ontario and eastern North America. Lakes and fish were already dying over hundreds of thousands of square miles when Harvey wrote his report. Geography and geology are part of the reason, particularly in Ontario. The soil is thin, lying on top of the million-year-old Precambrian rock which pokes up as the bare grey-pink slabs so familiar to cottagers, road-builders and geologists. The rock is hard, durable, and dominated by granite, quartz and iron. It doesn't wear down easily and it doesn't contain the elements which foster fertility, particularly calcium. It does not contain large amounts of chemicals to counter weak acids, to neutralize them as the acids flow through the thin soil and into the lakes and rivers. The land lacks, in the geologists' words, extra "buffering capacity."

It also means that thousands of years of erosion have yet

to deposit more than a limited amount of neutralizing chemicals into the lakes and rivers. The lakes are "soft water" lakes, they have low alkalinity. A typical lake with 500 units of alkalinity or "buffer" capacity can safely absorb five continuous years of fairly acid rain, such as pH 4.4, before its buffering capacity begins to decline. When the alkalinity is exhausted, the lake waters begin to turn acidic. But on the Precambrian Shield, from Georgian Bay to the Ottawa River, more than three-quarters of the lakes have far less than 500 units of alkalinity,[2] and the acid rainfall on the Shield has averaged less than pH 4.4, for more than a decade. Here, a geological-chemical balance which has functioned for 100,000 years is under great stress. "If recently recorded loadings of acidic materials to the ... lakes in the Haliburton area continue, then in many of these lakes the alkalinity will be exhausted in 5-10 years," an Ontario government report warned in mid 1979.[3] And with alkalinity gone, it takes only a small addition of acid to tip these lakes into a rapid slide down the acidity scale.

The implications of acid rain falling on weak soil can be seen in the toll of lakes already in trouble, or past the point of reclamation in central Ontario. Within 160 miles radius of Sudbury, for instance, 20 per cent of 209 almost randomly-selected lakes are "critically acidified," as the scientists say. The pH of the lakes is below 5 on average—it could run as low as 4, 1000 times more acid than neutral water — in the spring rush of acid shock loads. Those lakes are, in many cases, all but devoid of fish. Another 50 per cent are vulnerable, with low buffering capacity, limited and declining fish populations, and as little as 10 years remain until those lakes join the critical list. In that 3200 square mile area, nearly 15 per cent of the lakes have levels of toxic metals which exceed the government's criteria for successful survival of fish. In the 400 square miles of water surface in that area, the pH of the water has been dropping .1 units per year for at least the past decade. In other words, they are becoming 10 times more acidic every decade. The statistics make fascinating, if frightening reading for every cottager, tourist operator, environmentalist, and politician who cares about the area. The information is bound in a big green report the provincial environment ministry finished in 1977 (and quietly made public in 1978), titled *Extensive Monitoring of Lakes in the Greater Sudbury Area 1974-1976*.[4] The huge map bound into the report shows where the lakes are and how acidified they are. Most cottagers—most citizens—don't know it exists.

The Sudbury-centred study covers only a part of the province and a small sample of the damage. Nearly 1,000 miles northwest in a remote wilderness of Ontario along the Manitoba border, in what's called the Experimental Lake Area, 109 lakes have an average pH of 6.5 and falling. A few are worse. And in a typical month, July 1977 for example, the rain falling there averaged a pH of 4.5. But those lakes are remote; south in the cottage-crowded lakelands of central Ontario, within a few hours' drive of the urban core of the province, the lakes are acidified or weakening too. Near Dorset, for example, which sits on the boundary between Muskoka and Haliburton, conditions are similar, but the rainfall is even more acidic. During the winter of 1976-1977 snowfall near Dorset hit an unprecedented 2.97—the snow was as acid as vinegar. In fact, across the entire Precambrian Shield area of Ontario, in the kind of geography and geology least able to cope, the acidity of rainfall now averages pH 4, more than 10 times more acid than normal rain, 1000 times more acid than neutral water. The list of lakes receiving that rain stretches into the tens of thousands, the number already critically acidified exceeds 100, and hundreds or thousands more may reach the critical point in 10 to 20 years.

Ontario is not alone. Its neighboring province Quebec, with the exception of a narrow 50-mile-wide stretch along the St. Lawrence River, is almost exclusively Precambrian Shield. The rivers which flow into James and Hudson Bays average pH 6.2, and in the spring, 4 to 4.7. To the east on the Atlantic coast, particularly in Nova Scotia and Newfoundland, with their bedrock of hard granite and thin surface soils, the lakes and rivers are acidifying too. Interestingly, the decline there was charted as early as 1957 when biologist Eville Gorham measured 16 lakes not far from Halifax, and found they averaged about pH 6.5. Now 85 per cent of those lakes are 10 to 100 times more acidic than that, and worsening quickly, Their buffering capacity is nearly exhausted. Along the extreme southeast shore of Nova Scotia, federal fisheries biologist Walton Watt has nearly abandoned his search for Atlantic salmon in the nine major rivers which flow to the Atlantic. The Mersey, Roseway, Sissiboo and Tusket, all rivers which have swarmed with salmon for more than 200 years, flow empty now, the pH below 5. Salmon, like those Harold Harvey tried to introduce to Lumsden Lake in the Killarney Lakes of Ontario, cannot reproduce in even moderately acid waters. And with young fish lacking the instinct to return from the Atlantic, the salmon fishery disappears. Most of the trout are gone too. Only eels seem to survive.

Geology and weather know no political boundaries. In the Adirondacks of New York state, in the 6 million acre state park there, 170 of the lakes are acid dead, bearing no fish at all. And those 170 are a small sample of almost 2,000 more lakes to be tested. The Adirondacks is a one-day drive away from more than 55 million urban-dwellers, millions of whom use the park and the surrounding uplands for recreation, fishing and, where permitted, cottaging. The Adirondacks bear an uncanny resemblance to central Ontario. In 1929-1937 the upland Adirondack Lakes were virtually neutral waters—only 4 per cent had less than a pH of 5. Today, 51 per cent of all the lakes above 2,000 feet elevation have values less than pH 5—and 90 per cent of them are empty of fish life. The rain which falls on the Adirondacks averages pH 4.

Farther south in the Shenandoah and Blue Ridge Mountains, where the soil lies thin on top of weakly buffered, hard rocks, the alkalinity of rivers and lakes is low. In North Carolina, a state not previously known to be highly sensitive to acidification, 83 per cent of 42 surveyed counties was found to have weakly buffered, low alkalinity water—less than 200 units of capacity to cope with acid rain for any length of time.[5] And acid rain of pH 4-5 has now been detected. It may have been falling for years. Researchers are now investigating the fish population, fearing the worst.

There is a group of lakes near Jacksonville, Florida, with a pH of 4.7-5.5. A second group near Lake Okichobee has pH 6.5. "It is clearly a non-natural process. The soils are low in buffering capacity, being essentially completely sandy. ... And the acidity of the precipitation is in a smiliar range to the acidity of the lakes," University of Florida scientist Eric Edgerton told one of the authors at a Washington meeting in late 1979. "We haven't studied the fish yet." North again in Minnesota, almost 1,000 lakes have shown a decline in pH, and studies of fish mortalities are beginning to show the tell-tale signs: absence of trout, and the missing years of certain age classifications. Even the wideopen spaces out west are not immune: on the Como Creek watershed in the Indian Peaks wilderness which straddles the Continental Divide of the Rockies, researchers William Lewis and Michael Grant of the University of Colorado have charted increasing acidity. They've just begun their work.

Some of the most important foods for fish are the least tolerant of acidity. In Sweden, Norway and in the Adirondack Lakes it has been discovered that the common water-dwelling invertebrates such as mayflies and stoneflies disappear at an acidity of pH 6. Trout can survive that acidity for awhile but they must seek alternative foods. Freshwater shrimp, another common food for fish, are equally vulnerable. The list of aquatic insects sensitive to declining pH is growing almost as fast as the research can be completed. Clams, snails and crayfish disappear in waters acidified below pH 5.5, either because their shells are dissolved by the acid, or they simply cannot absorb enough essential minerals to form shells. Plankton are reduced, as acidity drops below pH 6. Even the bacteria and fungi which helps decompose dead material such as leaves and debris into digestible food for fish and the invertebrates have been found to suffer acid shocks. This acid death of bacteria and fungi is suspected as one cause of abnormally clear water, the tell-tale sign of an acidified lake. Without decomposition, litter and dead material accumulates on the lake bottom, instead of breaking up and spreading through the waters, to give them characteristic colors.

The smallest algae which live suspended in healthy lake waters feed off floating decomposed material. These algae or phytoplankton create new organic material for the entire food-chain in the water. But with acidification, these basic life producers are depleted. The lack of algae may also be a cause of the chilling clarity of acidified waters, although the impact of acid is also felt on the larger green growing things—sometimes to the plants' benefit, but often to the lake's ultimate harm. Lake Colden in the Adirondack Mountains, which has increased its acidity 10-fold in two decades, shows an abnormal, dense growth of sphagnum moss on its bottom, which did not exist 30 years ago. Woven into a tight barrier, this Astro-turf of moss and other abnormally flourishing algae seals off decomposing material from recirculation in the water, reducing the bacterial decomposition which would produce oxygen. A new oxygen-free environment has evolved, and within it, new bacteria which release carbon dioxide, methane and hydrogen sulphide from the sediment. Gasses erupt in great bubbles—the probable cause of the garbage-dump-like odors which rise from the surface of some acidified lakes during the warmest part of the year.

This process of decay, called oligotrophication, means that fewer of the ions of acid are neutralized by the depleted biological community so the acid can cause further degeneration of natural processes, which in turn are less capable of combatting the acid, and so on, in an accelerating process. All of this is happening in water which is not yet sufficiently acidified to be fatal to most species of fish.[6] But it is starving them and every living thing in the lake, while adding to the acidification.

Oligotrophication is a slow process, and many acid lakes have yet to lose their final buffering capacities. And yet researchers are finding many living lakes increasingly devoid of fish. The answer to this apparently "premature" acid death may lie with aluminum. The most common metal on earth, it normally serves as a glue for hundreds of other metals, binding them together into soil and rock. And as such, it is part of buffering mechanisms which have helped the land and the lakes withstand centuries of slightly acidic natural rain. But a mere three decades of abnormal rain have broken these bonds, and increased concentrations of aluminum wash into the lakes. In already acidified water, aluminum itself turns to a compound more acidic than can be indicated on the pH scale. Even at acid levels not normally harmful to most fish, aluminum in the water is toxic. As little as 0.2 milligrams of aluminum in a litre of water acidified to pH 5 is enough to burn the gills of fish. For more sensitive fish, like trout and salmon, much less acidity and that same amount of aluminum is fatal.[7] In the Adirondacks, Sweden and Norway, researchers measuring acid lakes routinely find high levels of aluminum too. In the spring, researchers don't even have to measure—they can see concentrations of aluminum washed from the land as a silver sheen on the lake surfaces. As Carl Schofield, a Cornell University biologist who has spent a decade monitoring aluminum build-up in the Adirondack Lakes puts it, "Because the poisonous aluminum run-off moves rapidly across the tops of the lakes in a relatively solid mass, fish can sometimes find refuge at the bottom of the deeper lakes with a good oxygen supply. But in the shallower lakes the fish are caught in a deadly trap. They have the choice of dying from lack of oxygen at the bottom or from acids and aluminum at the top."[8] In New York, Maine and Florida, scientists have found aluminum concentrated 20 to 150 times above normal in lake bottom sediment in acid lakes, locked in chemical combination with nutrients which normally float free.

Copper, cadmium, zinc and lead wash into the lakes in elevated concentrations too, leached from the soils by acid rain, or falling with the rain itself, to create their own hazards. Suspended in the water and absorbed into fish through the gills, these metals concentrate in the liver and tissue and blood. Diagnosis: toxic metal poisoning.

And, as every Canadian knows now, there's mercury in our lakes. Since 1977 the Ontario government has been publishing a list of lakes in Ontario where fishermen can no longer consume the fish they catch. As little as 1 part per million of mercury is considered too much in fish flesh consumed on any regular basis by pregnant women and children. It accumulates permanently in the human body, and anything more than 20 parts per billion can cause subtle nerve disorders, more obvious problems like tunnel vision and loss of balance, mental instability and slurred speech, and finally paralysis and death. A single tablespoon of mercury in a body of water covering a football field to a depth of 15 feet is enough to make fish in that water unsafe to eat. Acid rain is not alone responsible for the mercury. When the Ontario list of mercury-toxic lakes was first produced it was believed the listed lakes suffered mercury from industry, particularly the pulp and paper sector, which had dumped contaminated wastes in years past, especially in the intensely-polluted English-Wabigoon River area and Lake St. Clair. However the list has grown to several hundred names now, many of them far from any known source of mercury. The provincial list doesn't speculate on the cause of this increase, but the answer is almost certainly acid rain. Some of the mercury ends up in lakes as fall-out directly from air pollution from the same sources as sulphur dioxide. Coal in particular contains substantial amounts of mercury and more than 80,000 metric tons fall to earth around the globe each year.

Mercury also exists naturally in land and water, particularly in the Precambrian Shield where the rocks have a high natural mercury content. Normally this mercury would remain locked in chemical combinations in the rocks and water as it has for unknown centuries. But when the waters turn acid, mercury is eroded into the lakes, where it is turned into a free-floating menace.[9] Although the exact chemistry is now uncertain, the connection is clear. In Ontario many of the mercury-unsafe lakes match those with a high acidity level. In northwestern Quebec, where rivers and lakes average only slightly acidified conditions, fish contain mercury far above the acceptable limit. Low pH lakes like Cranberry and Stillwater in the Adirondacks contain small-mouth bass with abnormal levels of mercury, and in the Boundary Waters Canoe Area of northern Minnesota more than 1,000 lakes vulnerable to acid have elevated mercury in tested trout, walleye and pike. In Scandinavia, the combination of high acid levels and high mercury concentrations is commonplace.

Is there any way to cure the lakes, once they have become acid baths? Much effort and hope has been placed in finding the cure, but in fact, as with cancer, the answer lies in prevention, not reaction. At best the treatment may succeed in individual cases, or stave off an early end, but at worst it amounts to fighting one malignancy with another of injurious or limited potential. Shortly after Inco Ltd.'s smelter in Sudbury completed its 1,200-foot Superstack to waft away its air pollution, the Ontario environment ministry's Sudbury Environmental Study team began dumping lime in some local lakes. The idea was not new, since lime had been used in European hardwater lakes for a century to fertilize the water and boost fish populations. Near Sudbury the intention was to add enough lime, a base or alkaline material, to neutralize four acidified lakes in the hopes of fostering vegetation, aquatic and fish life. It was more than reclamation, it was a try at resurrection. Middle and Lohi Lakes were small, intensely acidified waters within 10 miles of the Inco smelter, typical of hundreds more in the region. They were surrounded by the exposed bedrock, with limited human access, high concentrations of toxic metals, and no fish. For much of the summer of 1973 and 1974, crews in protective clothing and face masks cruised up and down Middle Lake pouring lime and crushed limestone into the water from their powerboats. By the end of 1974 they had dumped nearly 38 tons of the white powder or wet sludge into the water. By 1975 there were glimmers of a happy ending. The acidity of Middle Lake had been reduced 100-fold; the concentrations of toxic nickel, copper, and zinc seemed milder as the limestone captured the metals and took them to the bottom. Lohi Lake required a repeat lime dosage since it was already slipping back into strong acidity, but in both lakes the populations of some vital water organisms, insects and bacteria were showing signs of improvement, and a basis for a re-established fishery. The researchers experimented with fertilizer to boost the water organisms, and added two new less seriously acidified lakes for a duplicate test. As Scandinavian experiments had shown, liming was not cheap—well over $230 per acre of lake surface—and it ran the risk of poisoning the lakes with other contaminants. It also seemed to have a very limited timespan of effectiveness, but the ministry pushed on.

In August, 1976, the ministry dumped 2,500 small bass into Middle Lake, the first step to a restored fishery. The acidity of the water was still well above normal rain, and only a bit below neutrality. Conditions appeared right. But by 1977 not one of the fish had survived. Lohi Lake, despite liming up to a near-neutral condition, lost 1,200 added brook trout within 4 months. Levels of copper in the lake had been rising again, and it was suspected the fish had died of copper poisoning. Undaunted, in 1977 the researchers returned to the Lakes with a fresh stock of trout and the ultimate in experimental schemes. They built plastic swimming pools beside a nearby neutral lake, and dumped all the trout in there first, to "acclimatize" them for what came next. The trout were well-fed and watered. None died. Then half their number was trucked to Middle and Lohi Lakes. To duplicate the conditions, the half destined for the neutral lake were taken for an equally long and bumpy ride over back roads, and then they too were placed in huge carefully-constructed submerged cages. At each lake the cages were lowered slowly underwater, to avoid sudden temperature changes. Scuba-equipped divers were on hand to feed the trout, remove any which died, and take regular water samples. At Middle and Lohi Lakes, the lakes "reclaimed by liming," the divers had short work. Within 24 hours the trout were swimming in confusion, within 48 hours some were dead. At the neutral lake used as comparison, the divers worked all summer feeding healthy fish in their wire cages.

At Middle and Lohi Lakes, the fish died of copper, zinc, and nickel poisoning, the toxic metals acting alone or in combination. The liming had failed to reduce the toxic metal concentrations in the sediment, and continued acid rainfall had brought new overdoses, washing in more metals from the land and sky. As a senior environment ministry scientist admitted a year later, "liming has at best very limited effectiveness. We lost those lakes and we'll probably have to write them off as dead forever." Liming lakes only moderately acidified is still under investigation by the scientists, but as the Swedish National Environmental Protection Board warned in March, 1979, the task of liming lakes at the exact moments of spring acid shock is almost insurmountable. Mid-summer liming is too late to save fish breeding and spawning in the spring. Even mid-summer liming would have to be repeated again and again, to save what few fish might survive the spring runoff.

As Gene Likens, professor at Cornell University and an expert on the Adirondack's acid-dead lakes puts it: "Liming appears to be only a temporary, short-term pseudo solution to the problem in natural eco-systems, and may cause secondary effects. It can be compared to taking morphine before you cut your leg off—it might ease the pain but you still bleed to death."[10] Giving our lakes big doses of lime won't ease the pain very much, and yet some optimists still believe, in the face of the evidence, that liming is the cure-all for thousands of dead and dying lakes. Even if it worked, dumping more chemicals into a polluted environment is Alice-in-Wonderland logic. Preventing industrial pollution is the only answer.

# 2
# Acid in The Air

About 11 ounces of sulphuric acid fall on every acre of southern Ontario now, each year. This rain is usually 10 times more acidic than normal rainfall, and often as much as 15 to 20 times worse. At Woodbridge on the edge of Toronto throughout July and August 1979 the rainfall averaged nearly 100 times more acid than normal rain. But normal rain, with its pH level of 5.6, is *now* an anachronism over eastern North America. It hasn't been regularly raining clean rain for more than two decades, it's been consistently raining acid. The Toronto measurements are comparable to those found widespread around the Lower Great Lakes, the Boston-Washington cities-belt, and even in the Adirondack Mountains and on the rocky coasts of Maritime Canada.

There has always been some acid in natural rain, caused by such gasses as the carbon dioxide from green plants which turns to extremely weak carbonic acid in air, with a pH of 5.6. And volcanoes, forest fires and sea spray contribute small amounts of sulphur dioxide which turn to sulphuric acid. But dust and airborne chemicals such as calcium and potassium have kept the acids relatively neutralized for thousands of years. Ice which originally fell as snow in Greenland only 180 years ago was still almost neutral when examined recently. But in the last 50 years, the basic and natural processes of nature have been altered, almost exclusively by industrialized societies' incessant demands for more energy and resources. And the fallout from this change comes from the sky as rain 10 to 100 times more acidic than normal rain.[1]

The Europeans have known about it for at least 20 years. Early in the 1950's at Sweden's instigation nearly 175 meterological stations were set up across western and northern Europe and their measurements of the chemicals in rainfall logged regularly and reported back to the participating countries. By 1960 the measurements indicated that the pH of rain was sliding downward from pH 6 in the mid 1950's to pH 5.5 or lower in 1960. And in a massive review in 1968 of a decade's measurements, Svante Oden of Sweden described something more. There had been an area of strong acid (pH 5) rain centred over southeastern Britain, northern France and the Benelux countries since 1956 or earlier. By 1966 the acidic area had expanded to include all of western and northern Europe, and in the original centre, pH 4 to 4.5 was common. A second centre of acidity over the western USSR and eastern Europe was also growing. By 1970 the pattern was firmly fixed and largely unchanging. De Bilt in the Netherlands recorded an annual pH of 3.7 in 1967, and on April 10 1974 Pitlochry Scotland reported the most acidic rainstorm ever measured in Europe— pH 2.4, equal to vinegar. Almost equally low pH values were detected at the same time in Norway and at a remote site in Iceland.

The impact of that acid rain on lakes and fish was already clear to the Europeans, particularly the Scandinavians, by the late 1960s, and the sources of the rain were also increasingly obvious: the tons of sulphur dioxide air pollution being ejected into the air from factories and power generating plants fueled by coal, throughout Europe. Sulphur dioxide air pollution had long been known and measured by European scientists, but was always considered a relatively short-term and localized nuisance. Before 1950 it was believed that the oxides lasted no more than 5 or 6 hours before apparently dissolving in the air, dissipating to immeasurable amounts or falling to earth. By 1963 however meterologists such as C.E. Junge—an American—confirmed that the pollution could last much longer, specifically up to seven days. By 1970 Oden and others had charted the airborne flow of sulphur dioxide from central

Europe to Scandinavia over five days. For example, Norwegian scientists calculated that 4,000 tons of sulphate and sulphuric acid fell on a 12,000 square mile area of Norway in an 8-day period. It came from central and western Europe.

By 1973 the researchers had counted up 25 million tons of annual pollution coming particularly from central Britain, the Ruhr Valley, the coal fields of East Germany, southern Poland and Czechoslovakia. As had become clear, the pollution emissions accumulated in the atmosphere faster than they could dissipate or fall back to earth. They were overcoming the natural buffering chemicals of the air, and falling as acid rain or as particles of sulphate wherever the air chose to move them. In fact, more sulphur dioxide was being put into the atmosphere than was falling back out. Hence the gradual build-up of permanently acidified masses of air hanging over half the continent. Sweden slapped reduced sulphur dioxide emission limits on its industries in 1969 and 1976, and began protracted negotiations with upwind neighboring countries to do the same. Britain, having suffered thousands of deaths in the 1950s due to the Killer Fogs laden with sulphur dioxide, cut back uncontrolled use of coal for fuel by 1964, and introduced wider use of natural gas.

The record of acid rain in North America is much less clear, although the impacts are potentially worse than those in Europe and Scandinavia. Even today the network for monitoring acid rainfall on this continent is inadequate to give a detailed picture. But it is known that before 1930, our rainfall was clean, at least in New York, Virginia and Tennessee where the earliest recorded samples were taken. In 1939 a measurement of a Maine rainstorm produced a value of 5.9., but by 1955 the rain had turned a bit sour. Maine to New Jersey was often under a cloud of rain 10 times more acidic than normal, and that cloud stretched all the way west to Ohio. By 1966 the New York to New England centre of airborne acidity had dropped to pH 4.4 and the nearest area of regularly clean rain was South Carolina.

Strangely, the US government sampling network which

had charted these changes since 1959 was discontinued in 1966. But based on university, individual state and experimental federal samplings, the 1973 picture was calculated to be even worse. Nowhere east of the Mississippi River was there an annual average of clean rain, except for the southernmost tip of Florida. Every area north of Tennessee suffered rain at least 10 times more acidic than clean rain. And in upper New York and Connecticut, the average rainfall was 50 times more acidic than its clean levels of 35 years earlier. Those conditions generally prevail today, with one exception. The area of 10-fold-more-acid-than-clean precipitation is still spreading. Somewhere in the range of pH 4.5-4.0, acidity in the atmosphere hits a temporary plateau—it takes a lot more sulphur dioxide to push the pH a small amount lower than it does to push it from 5.6 to 4.6. But like an oil-spill on an ocean, instead of consolidating in a small puddle, acidity spreads further at an optimum speed and depth. From the northern states, the acid pall is spreading south and west at a pH of 4.5 by hundreds of miles every 5 to 10 years.

Although sulphur dioxide is the most prominent pollutant in acid rain, monitors at the Hubbard Brook Station in New Hampshire have noted the presence of other chemicals, such as potassium, magnesium and nitrogen. The Hubbard Brook scientists calculated that nitrogen oxides were probably responsible for 15 per cent of the acidity in the rainfall: nitrogen oxides, like sulphur dioxide, can be transformed to weak nitric acid in air. Decaying plants, bacterial action, and lightning are prime sources of nitrogen, which exists both in soil and vegetation, and in air itself. The Hubbard Brook analyses revealed a startling change from 1964 to 1974: although the acidity of the rainfall had become only slightly worse, the percentage of sulphuric acid had dropped and the percentage of nitric acid had doubled: it now contributed one-third of the acid. Although it took four more years to confirm, the scientists at Hubbard Brook had just discovered the other major acid in acid rain, and had charted its ominous growth. They were largely unaware that Oden of Sweden had detected similar signs in Europe and Scandinavia in 1968.

By 1978, the data had begun to accumulate: nitrogen oxides were a substantial and growing contributor to acidity. In some places as remote as Pasadena, California where the rainfall was average pH 4, nitric acid was the dominant component. Pasadena had no ore smelters and is not consistently downwind from major polluted air masses. Its local and regional fossil-fueled power plants burn western US coal which is noticeably lower in sulphur. But Pasadena has cars and trucks. Nitric acid is about 30 per cent of the cause of acid-rain in North America; transportation is nearly 50 per cent of the source. The other 50 per cent comes from combustion of fossil fuels, coal in particular. In Canada, cars and trucks contribute over 60 per cent of the nitrogen oxides, power plants about 10 per cent, and other industrial combustion about 20 per cent. But not the ore smelters like Inco of Sudbury. The non-ferrous smelters produce negligible amounts of nitrogen oxides. In the US transportation produces 40 per cent, electric utilities 30 per cent, and other sources 20 per cent. The nitrogen oxides are formed when the intense heat of combustion causes some of the natural nitrogen in all air (78 per cent) to burn as well, turning it into oxides. About 24 million tons of nitrogen oxides were fired into the air in the US in 1978; about 2.1 million tons in Canada.

As with sulphur dioxide, two-thirds of the sources of nitric acid rain are located east of the Mississippi and the Manitoba border in North America. And with certain exceptions, once the nitrogen oxides are up the stack or out the exhaust pipe they behave like sulphur oxides, slowly turning to acid in air, rapidly blowing hundreds of miles across the continent, and eventually coming down as acid. Not all the sources of nitrogen oxides have been clarified, and there is increasing evidence that heavy use of nitrogen fertilizer may be a source of nitrogen oxides. There is however clear evidence that nitrogen oxides have been increasing at a much greater rate (from 11 million annual tons in 1950 to 26 million in 1975) than sulphur oxides in North America, largely due to the continued rise in the number of automobiles. The catalytic converters now standard on new vehicles purge carbon monoxide and hydrocarbons from the fumes but are not nearly as effective (yet) against nitrogen oxides. And finally, nitrogen oxides react with ozone and some hydrocarbons in the presence of sunlight to form photochemical smog, the kind of yellow-grey haze which is literally alive and growing in stagnant air masses over areas like Los Angeles, New York, Washington, and southern Ontario, and carries serious risks of respiratory irritation and crop damage. Every year since 1976 federal Atmospheric Environment Service scientists in Toronto[2] have charted the flow of ozone-laden air masses sweeping in from the mid-western US across the Great Lakes at levels as high as 200 parts per million—more than twice the Canadian maximum tolerable level, thousands of times higher than the maximum 30-minute permissible US level. Twenty miles northwest of Toronto, having combined with nitrogen oxides emitted by Toronto, the ozone turns into a smog which is etching holes in tobacco leaves, and holds still unstudied risks for human health.

When US scientists like Gene Likens and C.V. Cogbill published their startling reports on the spread of acid rain, in 1974 and 1976, they also stretched their maps north to include Canada. No area south of Thunder Bay to Newfoundland enjoyed "clean" rain, and in a circle of 30 miles radius around Sudbury, the rain measured pH 4.23, which made this spot in northern Ontario the second most acidic area on the continent. The Sudbury acid rain measurements came from Canadian studies done in the giant smelter area as early as 1960, and from post-1970 work intent on showing that the erection of the 1,200 foot Superstack in 1972 would lead to cleaner air for the beleaguered smelter city. As early as 1970 the federal Atmospheric Environment Service had totalled up the annual amount of sulphur dioxide going into Canadian skies—7.2 million tons, including the principal sources—smelters, industries and refineries, and power plants.

It would have been logical to study what such a poisonous volume of air pollution was doing to the atmosphere, but governments, especially regional ones, seemed reluctant to take that next step. Not until 1976 did two scientists with the federal Atmospheric Environment Service first published an assessment[3], in a little-read technical journal, which showed the inevitable similarities to the US acid phenomenon: across southern Ontario the rain averaged pH 5 to 4 and lower, and rains even more acidic had been detected. Quebec was an unknown but the Maritimes was already below the "clean" level of 5.6.

The federal Service stepped up its eastern Canadian air monitoring network and by 1976 was recording an increasingly acidic rain across southern Ontario. That same year the provincial environment ministry began taking some interest in the acidity of rain beyond the Sudbury smelter area, and in the first publicized ministry report in October 1977 the reality of the sulphur-laden atmosphere over Ontario was acknowledged. More than 180 miles southeast of Sudbury at Dorset, Carnarvon, and near Bracebridge, the rains were 10 to 50 times more acidic than normal "clean" rain. At Dorset in the winter of 1976, a snowfall measured 2.97 on the pH scale. These measurements came nearly five years after Harold Harvey first warned that airborne sulphur dioxide meant long-distance acid rain, and three years after the Swedish experts had warned Canadians at their Winnipeg meeting to look further afield for the telltale acidic signs. In the mid 1970's in Canada, particularly in Ontario, there had been many more experts in the field charting the death of lakes and fish by acidity than there were those measuring the rain's pH and the range of its devastation. It seemed to take Canada a long time to make a connection between obviously massive outpourings of pollution and the abilities of regional air reservoirs to sustain contamination without change. In Canada, the official federal listing of sulphur dioxide sources was not published until *1977*! In their slow and stately response to the problem, both Canada and the States seemed determined to follow the grand old tradition already set down by their earlier attitude toward water pollution: "out of sight, out of mind."

Unfortunately, there is no way to hide acid rain. Today in Canada the once-pure air reservoirs are also vats of sulphur dioxide which spill acid back on the land to blight crops and forests. In the recreational wilderness of Ontario the rainfall averages pH 4.1-4.3. Further south, off the Precambrian Shield and into the urban and agricultural heartland of the province, federal measurements plot the average pH at 3.9 to 4.1, which equals the worst levels in the US, and much worse than levels in Scandinavia. As late as 1979 the Ontario government had no acid rain monitors near the farms and cities. There was no need, claimed the environment ministry, because acid rain posed no risks to those areas. (The Toronto measurements were taken by a federal monitor.) As federal monitors elsewhere show, nothing east of Manitoba and south of James Bay enjoys clean rain on an annual average, in Ontario. Quebec through Nova Scotia averages 4.5—10 times more acid than clean rain. The only area which has shown any improvement is the tiny circle around Sudbury, Ontario, almost within the shadow of the world's tallest smoke stack. The fumes from the 1,200-foot Superstack blow elsewhere.

The Sudbury Superstack of International Nickel Ltd. (Inco) is also the world's greatest single source of sulphur dioxide, since 1978 licensed by the Ontario government to emit 3,600 tons per day, 1.3 million tons per year, 1 per cent of all the natural and man-made sources in the world, and equal in the past decade to all the emissions of all the volcanoes around the globe. In earlier decades the smelter emitted far greater amounts. The sulphur dioxide comes from millions of tons of rock mined beneath the city of Sudbury and smelted in a process virtually unchanged in eight decades. Five thousand tons of rock are crushed each day and melted in enormous cauldrons; then the nickel-bearing ores are separated and burned again to free the nickel for recovery. The left-over slag is dumped in 200-foot-high ridges which stretch miles from the smelter, and the 3,600 tons of sulphur dioxide burned out of the rock and ore goes up the stack. Additional tons of sulphur dioxide escape at ground level, through the walls, doors and ceilings of a sprawling smelter complex which stretches for city blocks, and can be seen and smelled for miles.

The Inco smelter is visibly the largest air polluter in Canada, by virtue of its Superstack, its pollution plume, and its emissions, which make up 20 per cent of Canada's 1978 total of 5.5 million tons of sulphur dioxide per year. But Inco is not alone. Together with other smaller smelters such as those at neighboring Falconbridge, and Rouyn-Noranda, Quebec, Thompson, Manitoba, and Trail, British Columbia, the non-ferrous smelting industry is responsible for almost half of the country's total emissions. The smelters dominate the national economy and the national air resources. They emit twice as much sulphur dioxide as all other industrial processes combined. Transportation, electrical power generation and other sources of fossil-fuel combustion make up the remaining 25 per cent. In Canada in 1970 there were only 43 sources of sulphur dioxide emitting more than 3,000 tons of sulphur dioxide per year including ubiquitous urban emissions. The smelters' emissions dominated the government pollution production charts like smokestacks dominating a Sudbury horizon. Nothing has changed today. Ontario and Quebec, with the largest smelters, produce 60 per cent of the national sulphur dioxide emissions total.

South of the border, the total American emissions of sulphur dioxide are five times greater than the Canadian—28 million tons in 1978 versus 5.5 million, and two-thirds of this comes from electric power plants, versus one-sixth in Canada. Non-ferrous smelters in the US emit almost the same volume as Canadian ones, but on the American scale they make up less then 15 per cent of the total. Never as blessed with fast rivers to dam for hydro power, and burdened by a 10 times greater demand for electrical energy, the US has long relied on burning coal (and oil) to turn water to steam to drive the generators. Coal already provides 50 per cent of US electricity. No single coal-powered electric plant comes close to the 1.3 million tons per year emission of Inco in Sudbury, Ontario. But taken as a single company with several sources of air pollution, the Tennessee Valley Authority with its 14 power plants in the eastern US is the worst corporate polluter on the continent, putting out 2.5 million tons per year in 1975. Inco came second.

In 1975 for example, more than 600 million tons of coal was stripped and mined from US fields, most of it from Ohio, Pennsylvania and the Appalachian Mountains which bisect the eastern half of the country. And where there's coal, there are power plants, at least 284 of them east of the Mississippi, but particularly concentrated in the Ohio River Valley states of Ohio, Indiana, and Kentucky. Coal, like the nickel ores of Sudbury, is laden with sulphur, and from the 284 plants rises nearly 80 per cent of the 18 million tons of sulphur dioxide from power plants which fills American skies and turns to acid rain. The plants are jammed together around the cities and steel smelters, some of them emitting more than 800 tons of sulphur dioxide per day, as at the Tennessee Valley Authority's plant in Muhlenberg, Kentucky, the Ohio Power Corporation at Morgan, Ohio, or Union Electric at Franklin, Missouri. Some of them have been operating for 40 years, some at even higher emission rates than they manage at present.

Regardless of the differences in sources and volume, the pollution process is the same on both sides of the border. Smelted out of the ores or cracked free from the molecular structure of the blazing coal, sulphur combines with oxygen to form sulphur dioxide and is carried skyward in a convection of ashes and other chemicals, (including, from coal, mercury and radioactive elements.) By itself, the sulphur dioxide gas can damage vegetation and property, and is a known respiratory irritant. But pushed upwards, and subjected to humidity, temperature, the presence of neutralizing chemicals in the same air current, and by sunlight, some of the sulphur dioxide begins to fall back to earth. Vastly more of it remains aloft and begins to transform from sulphite to sulphate to sulphuric acid, in combination with the inevitable molecules of water found everywhere in the atmosphere. At high humidity, up to 55 per cent of the total sulphur can be turned to sulphuric acid in two hours. In a wind of only 11 miles per hour, the same acid can be carried 60 miles before it falls. And at lower humidity, both the sulphur dioxide and the sulphuric acid remains airborne even longer.

In an atmosphere already heavily loaded with pollutants which have already absorbed what neutralizing chemicals there are, the sulphur dioxide coming from a smokestack literally joins the clouds and after a few hours can stay suspended for five days or more. What was once a local air-pollution problem is on its way to becoming a continental one. Eventually, the sulphur dioxide returns to earth, sometimes caught and washed out of the clouds by rain falling from above or simply carried downward as the clouds themselves condense into rain droplets, Some even falls in a dry form, as microscopic particles, to soak into the land in search of the moisture to turn it to acid. Sulphuric acid measured at a pH of 4 is only 0.0005 percent acid by weight in air, but over a year, the microscopic quantities add up: 11 ounces of sulphuric acid fall on every acre of southern Ontario per year now, and another 44 pounds come down in the dry form. As Dr. Marie Sanderson, a geographer at the University of Windsor, commented in 1977 after she first worked out the 11 ounce acid-fall calculation: "There's a continual supply of it up there. Every time it rains more pollution is coming down to earth." And where it falls, exactly, nobody can predict.

The scientists call acid rain's striking power "the long range transport of atmospheric pollutants." In simple terms this means that most of our acid rain blows in from somewhere else. Ironically, the continental spread of airborne acid is partially the result of shortsighted attempts to clean the air around localized sources of the pollution. Between 1955 and 1965, major sulphur dioxide air polluters embarked on a crash campaign to end years of neighbors' complaints about low-lying sulphur dioxide gas, dust and ash emitted from the coal- and oil-fired boilers and engines. The cure-all, seized upon just as the US government began imposing new standards to protect the quality of ambient air, was to build tall smoke stacks. Fumes emitted at a higher altitude would blow farther away and be continuously diluted and dissipated to fall eventually in apparently harmless, almost undetectable concentrations. In 1955 there were only two smokestacks in the entire US taller than 600 feet. Today at the 284 coal-fueled power plants in the eastern US, almost every stack is 600 feet or taller. Some soar to more than 1,000 feet. And yes, local ambient air quality has improved in some areas.

Meteorologists and air pollution experts can be pardoned for their endorsement of tall stacks back in the 1950s. At that time it was still generally believed that sulphur dioxide couldn't last more than 5 or 6 hours in the air without dissolving into relatively neutral elements. But by the mid 1960s the Scandinavians, already aware of acidifying lakes, had confirmed both the long-term stability and long distance transport of acid pollution from the stacks of Europe. By 1971 they had developed sophisticated models of what they called "long distance dispersion." And in August 1974, the Norwegian Institute for Air Research sent a pollution-sniffing airplane aloft into the path of winds blowing in from Europe.[4] When two days of flight data were analyzed, the pattern was clear. On August 26, thousands of tons of sulphur dioxide were detected rising in the air over Germany, Poland and England. Wind currents were blowing from the continent and England in a narrowing trajectory over the southern and western coasts of Norway. On August 27, as the dioxide-laden winds encountered Norway's coastal mountains, the air masses rose upwards, cooled and then condensed into rain. Nearly 1.5 inches of rain fell that day, along the coast and inland, and it was highly acidic—as low as pH 3.6, 100 times more acidic than normal rainfall. By the end of the day the European air mass, somewhat cleaner now, had travelled over 500 miles into Norway, more than 1,000 miles from its source, and it was still acidic. It's worth noting that Svante Oden, the Swedish grand master of acid rain studies, had already predicted the pattern as early as 1968.

The same kind of process has been happening in North America for at least 20 years. Blown eastward by prevailing winds, the mass of sulphur dioxide from the coal-fired plants of the Ohio Valley rises as it meets the Appalachian and Adirondack Mountains, then cools and condenses to acid rain. Farther to the northeast, this air mass meets cooler northern air over Ontario, Quebec and the Atlantic coast and more rain and more acid falls. On other occasions, dioxide-laden currents from Ontario and Quebec sources push south and east into the US, to encounter warmer air there and again condense into rain, acid rain. This continental wind pattern had been mapped out for years in North America, but even as late as 1971 the "tall stack" cure for local air pollution was still seen as the solution. This was particularly true in Ontario where the provincial government had fully endorsed Inco's proposal to build its Sudbury Superstack.

The claims of Inco official R.R. Saddington in the Toronto *Globe and Mail* on July 26, 1971, on the same page as Harold Harvey's first warning of acid rain damage in the Killarneys, reveal something of the ignorance—or deception—which beset the thinking in Ontario then: "Fears have been expressed that the tall stack will simply spread pollution over a wider area and put more poisonous sulphur dioxide in the atmosphere. In fact, it will do neither. There is widespread misunderstanding of the nature of sulphur dioxide. On a worldwide basis a full 80 per cent of the sulphur dioxide in the atmosphere comes from organic decay. About 14 per cent of it can be attributed to the burning of fossil fuels and about 6 per cent to smelting operations." Saddington went on to say that "Sulphur dioxide survives only about four days in the lower atmosphere. It does not accumulate in the air as a poisonous layer in the earth's atmosphere. Therefore, the problem is not so much the volume disseminated from a stack but the ground level concentrations." So why worry?

Inco is a multinational corporation with experts in every field, including the environment, the company says. But the experts must have been out to lunch the day Saddington wrote his praise of Superstack. Because the Superstack boosts pollution higher and farther, less of the air mass encounters the ground right away in order to drop its acidic particles. The wider dispersal means that more pollution stays airborne longer. So the Superstack could "safely" pump more pollution into the atmosphere without any noticeable local effect. The long-distance effect was another, unconsidered, matter. As for the natural sources of sulphur dioxide vastly outnumbering the man-made ones like Inco, such well-known experts as Junge had calculated as early as 1960 that in the industrialized northern hemisphere, man-made sources were at least equal to natural sources.[5] Could Inco's experts really have told Saddington that sulphur dioxide sits still in the atmosphere and dissolves in four days? By 1968 the Scandinavians had been working for a decade with air-mass trajectories, conclusively showing that four days gave sulphur-laden air plenty of time to take a long trip and dump tons of acid elsewhere.

Saddington was right about one thing. Sulphur dioxide doesn't accumulate in a poisonous layer. It moves away in a poisonous mass, but this fact did not concern Inco or the provincial government, who wanted some relief for Sudbury. Even Sudburians were getting tired of living in a moonscape of stripped backyards and dwarfed trees, and Superstack, completed in 1972, relieved pollution at the local ground level. Five years later, early in the morning of August 30, 1976, scientists Millan Millan and Y.S. Chung at the federal Atmospheric Environment Service on the edge of Toronto switched on instruments designed to measure high altitude sulphur dioxide.[6] At 11:30 that morning the needles on the dials began to jump and by 1:30 in the afternoon they constantly registered an abnormal mass of sulphur-laden air 1,000 feet overhead, with sulphur dioxide concentrations three times above normal. The polluted "air parcel," as the scientists call it, lasted until late afternoon. Within a matter of days, using detailed weather maps, measurements and regional reports the two scientists had tracked the pollution 250 miles back to its source: Inco's Superstack in Sudbury. As they learned, on August 29, an air mass of light northerly winds had begun pushing the stack's air pollution directly southward (the stack was emitting 15 pounds of sulphur dioxide per second).By mid-day on the 30th, the southernmost tip of the pollution plume was over Toronto with about 17% less sulphur dioxide. The 4-mile-wide plume wandered back and forth over a 30 mile-wide zone between Toronto and Oshawa during the afternoon of August 30, and when last measured was heading southeast across Lake Ontario towards Rochester. The plume had passed directly over the Killarney Lakes, Wasaga Beach and Orangeville, on its way to Toronto. The two scientists wrote up their report, later published it, and added the data to their files, the files including the color picture taken September 1, 1974, which showed the Inco plume stretching across the sunset sky for a distance of 50 miles, and the satellite photo taken on September 9, 1972, showing an Inco plume stretching 150 miles southwest to the Bruce Peninsula. And those were the days when the weather was clear, when it wasn't raining.

Admittedly, those three days of Inco plumes were exceptional. Under normal fluctuating weather conditions which sweep Ontario, Inco's pollution is much less easily tracked, and in fact often pours into an Ontario sky already laden with imported US sulphur dioxide. The back-and-forth flow of acid rain across the international border has become a most intensely debated scientific and political issue, and this so-called "trans-boundary flux" is one of the thorniest problems now facing Canada's relations with the United States.

Resolution of the debate in absolutely detailed scientific terms is now being used as a tactic to stall any political action on either side of the border towards the only issue which really matters: turning off the acid rain at its source, regardless of borders. Nonetheless, the facts and estimates of transboundary flux of acid rain reveal much about the enormity of the problem facing Canada and the United States. They also confirm why both countries must clean up their air pollution act.

Across the middle of North America, the prevailing winds are west to east and then to the northeast. Such winds are the prime cause of much of 15 million tons of sulphur dioxide emitted from 14 US states east of the Mississippi ending up as acid rain over the Adirondacks and Appalachians, and the Atlantic seaboard. For example, between August 9, 1976 and August 14 sulphate levels over Halifax, Nova Scotia rose 18-fold.[7] The polluted air mass, tracked backward, had begun its flow nearly five days earlier over the coal-fired Ohio Valley power plants. A typical 12 miles-per-hour wind can carry a mass of air 870 miles—from Chicago to Montreal—in as little as three days. And in that three days as much as half the sulphur dioxide has been transformed to acid rain droplets, ready to fall. But North American winds do not always blow so consistently; there are important seasonal variations: winter winds frequently blow southward from Canada and then southeast across the US, bearing relatively clean air (with the exception of Inco's emissions, among others), while in summer slow-moving air currents push north from the Gulf of Mexico picking up and dropping pollution as far as northern Quebec. Amidst these dominant air patterns, acid rain is being exchanged across borders, and the exchange is not necessarily equal. With a five times greater total emission of sulphur dioxide, the US obviously has a lot more pollution to offer, and the wind does blow northeast towards Canada much of the time. In late 1979 both federal governments released their first somewhat cooperative effort to sort out the flux[8], and with surprising results. Both countries' studies indicated that on a national basis the US is dumping three to four times as much airborne sulphur into Canada as vice versa. (Admittedly a far cry from the six to one ratio Ontario's environment ministry had previously claimed.) Studies of how much sulphur actually fell to ground, as dry particles or wet (acid rain), in the critical northeastern region produced similarities too: both agreed the damage lay between 5.4 and 5.8 million metric tons of sulphur annually falling on the northeastern US, from whatever sour-

ces. The Canadian researchers also produced a calculation that
about 4.2 million metric tons were falling on eastern Canada,
from whatever sources, a figure surprisingly close to the US
total, considering the vast differences between Canadian and
US industry.

US scientists calculated that in two typical months, January and August, 1977, twice as much sulphur fell on northeastern Canada as was sent south to the US by Canadian sources, a
finding which came as no surprise to the Canadians. However,
the US study went on to calculate that in those two typical
months, nearly 105,000 tons of sulphur produced in Quebec
and Ontario fell right back down on Ontario and Quebec.
Patriotically, Canada was keeping its home-grown pollution at
home, and more of it was ending up in Canada's backyard than
came from south of the border. It was an unsettling finding,
and Ontario in particular took exception to the home-grown,
home-sown analysis, having six months earlier argued that US
sources were by far the largest contributor to Ontario's acid
rain. Ontario environment ministry studies showed that for 3
months, while Inco's Superstack was shut down in 1979, acid
rainfall was unchanged in Ontario, so the Americans were
obviously to blame. This highly publicized study dealt only
with a 60 mile radius around Sudbury, and it was done in the
summer when the winds traditionally blow from the south.
And it neglected to say where Inco's 3,600 tons-per-day pollution was going every day when the smelter was in operation.
However, the study became an important support for Ontario's reluctance to lean too hard, too soon, on Inco. And by
dealing only with Inco and the north-central part of the province, it fully ignored the southern portion where Ontario
Hydro belched its own sulphur dioxide emissions.

On both sides of the border, scientists responsible for the
first joint Canada-US study of transboundary flux went back
to their charts and computers to fine-tune their models, and
fudge some more factors. But it was obvious what the first
study had produced and the next round, scheduled for late
1980, would confirm. Both countries are seriously polluting

each other with acid rain. Both countries have to stop. This, of course, involves political decisions and is outside the realm of science. The scientists preferred to measure the minute, and let someone else worry about the obvious.

# 3
# Acid in the Ecosystem

The Hubbard Brook Forest of New Hampshire stands against the White Mountains, thick with sugar maple, yellow birch and beech trees. Red spruce and mountain maple are scattered here and there, outposts of the boreal forest which begins further north and stretches far into Canada. Between the cool, moist summers and the deep snows of winter the slopes blaze yellow and orange, and crimson sumach accents the outcroppings of bare granite. The last loggers moved on in 1917. The forests and thin soils have never suffered fire damage and trees predating the first settlers in 1770 still remain, their 30-inch-diameter trunks too twisted for lumber yards. The majority of trees are 80 to 100 years old, and have formed a thick green canopy 30 to 50 feet above the ground.[1]

Between 1968 and 1972 this natural preserve was invaded by man. In the previous decade, foresters and botanists had wandered the forest snipping leaf cuttings, scratching for bark samples, and counting arboreal species. But in 1968 a new crew arrived, and soon the forest buzzed with the sound of chainsaws and steel drills. It was part of a painstaking scientific study by Yale and Cornell Universities, the US Forest Service, and the Brookhaven National Laboratory, a national research agency. Almost 100 trees were sacrificed, representing every major species, classification and location. The trunks were sliced and diced like carrots, the bark peeled, cores drilled, branches weighed and measured, germinating seeds counted, and the roots, "excavated with the encouragement of dynamite," as a later study described it, were boiled, powdered and microscopically analyzed. Even branches which had fallen during the winter were collected. Data from another 497 trees were added to calculations fed into the computers.

It took nearly four years to pull together the findings. To 1950 the forest had grown in a normal fashion; between 1961 and 1965 the production of wood in the trees abruptly, sharply decreased by nearly 20 per cent. In two centuries of previous growth nothing like this decrease had ever happened. But between 1961 and 1965 acid rainfall had established itself over the northeastern US, including the White Mountains. In 1974, the scientists reported that "increasing acidity of rainfall may be responsible for the decrease in forest productivity," and suggested, with academic caution, that the decrease of forest function through pollution was "deserving of further study." Before the decade's end, further study certainly seemed to confirm what no one would think of doubting: acid-soaked rain and acid air are not good for plant life.

In the early '70s, five hundred and sixty miles northwest of Hubbard Brook Forest, in the Sudbury, Ontario, region a pair of Canadian scientists were picking their way through what had once been a mixed boreal forest. Only white pine and jack pine, red maple and red oak trees remained, widely-spaced, often stunted, their foliage sparse. In the 30-mile radius covered by scientists Hutchison and Whitby, much of the terrain was bare, blackened granite rock. At only nine sites did they find enough dirt to take samples at carefully separated depths. They measured the soil's acidity when wet and dry, bagged samples for later analysis, and where possible, scooped up small boxes of the soil intact. They checked air monitors near these sites to measure the acidity and composition of the wind which blew over the area, particularly from the Inco and Falconbridge smelters, the centre of their research zone. Within a few miles of the smelters they already knew what to expect—sparse soil heavily contaminated with toxic metals, parched of essential alkalinity and dosed with sulphuric acid.

As early as 1954 such scientists as Sam Linzon, then

working for the Ontario department of lands and forests, had linked the death of trees directly to the influence of fumes which had blown for decades from the smelters in high-concentration, ground-level masses of sulphur dioxide, before taller smelter stacks had been erected. But for Hutchison and Whitby, the samples taken as much as 30 miles downwind held new surprises. After nearly two years work, including efforts to grow new plants rooted in the Sudbury soil under clean-air conditions, they concluded that most of the soils were hopelessly poisoned with metals freed from natural bonds by Sudbury's acidic rainfall. Organic matter needed for resisting excessive microbial degradation and for transmitting nutrients was missing; and the soils were loaded with sulphuric acid. They noted that the local air around Sudbury was noticeably cleaner after the Superstack had been erected in 1972 and conceded that much of the destruction could be due to previously unstudied effects of past air pollution. But "the disturbing implications would be if these (soil and vegetation) effects were to be repeated on even wider scales" in areas with similar original soil and acid rain levels. "The prospect of increasingly widespread acidic rainfall over hundreds of square miles (due to the Superstack) is not one to be taken lightly," they concluded in their reports issued in 1973, 1974 and 1977.[2]

While American scientists were dynamiting the roots of Hubbard Brook Forest, and Hutchison and Whitby plucked sparse vegetation from Sudbury's moonscape, Sweden released its massive acid rain report to the United Nations' Stockholm environmental conference.

The impact of long-distance acid rain was indisputably confirmed by two decades of Swedish measurements. Based on samples taken from more than 200 locations across Sweden and Norway, 100-mile-wide swaths of land were noted with abnormally low pH values, weak buffering-alkalinity capacity, and low proportions of essential buffering chemicals, The areas were largest where the fall of acid rain was heaviest—along the southwest, south and southeast coasts facing Britain and Europe where nearly 20 million tons of sulphur dioxide was generated each year and wafted towards Scandinavia. The soils were often thin, dominated by sandy material eroded from the underlying granite bedrock, with limited depth of decomposed humus and high concentrations of heavy metals. The Swedes calculated that these soils held only 5 to 20 per cent of the acid-resistant basic materials they might have held under less acid-influenced rainfall. And they calculated that the soils were losing their alkalinity by as much as .1 per cent per year, under continued acid rainfall. This, they predicted, would have serious effects on the productivity of the soils, and vegetation.

Soil productivity and growth is an extremely complicated and subtle process, the Scandinavians conceded, so much so that increased acidity of soil might take decades to produce visible changes in vegetation growth, "when it is too late to do anything about it." So the Swedes went looking for what had already happened.[3] They sectioned southern Sweden into two areas, one of strong acid rainfall, acidified lakes and rivers, and Precambrian soils of typical low buffering capacity, and another area of similar rainfall but deeper, more neutral soils. Researchers bored holes in 4,200 trees in the two areas, including trees 120 years old, and measured wood-fibre growth year per year. They found that somewhere after 1951 growth rates began to decline in the acid-sensitive section of Sweden. By 1965 the growth rate was an average of .3 per cent per year lower in the acid-susceptible area. Acidity of rainfall over Sweden had begun a sharp increase in the 1950s, to reach annual average of pH 4 or less by 1965 in some areas. "We have found no other reason for attributing the reduction in tree growth to any cause other than acidification," one team of researchers concluded. The change in acidity of the rainfall was easily charted; the change in forest growth was slower and harder to detect. But it was happening. The Swedish report to the UN Stockholm conference in 1972 also noted striking similarities between acid rainfall, soil composition and vegetation in susceptible areas of Scandinavia and in northeastern North America.

From 1972 to 1975 the Swedes, Norwegians and a handful of North Americans continued their drilling and sampling. The search for links between acid rain and soil and vegetation damage was vastly more complicated than rain-lakes-fish studies. Tree species vary widely in the kinds of soil they prefer, adjacent trees exert their own influence on the soil, and soil can vary from loam to sand within a distance of a few yards. Climatic variations, rainfall, temperature, elevation, exposure to sunlight, and even insect and bacterial activity make predictions of acid susceptibility and impact no easy matter. And by 1975 Scandinavian research discovered that the presence of nitrogen in nitric acid rain could actually serve as a beneficial fertilizer for soils and vegetation. Prior to the Swedish studies started in 1969, none of these factors had been considered. But in May, 1975, nearly 300 scientists from 12 countries including Canada and the United States met in Columbus, Ohio, for the first international conference on acid rain and the forest ecosystem.[4] Swedish experts such as Svante Oden outlined the spread of acid rain in Scandinavia, Norway's $1.5 million-per-year acid rain study by 40 full-time scientists was reviewed, long-range North American transport of sulphur dioxide pollution and its impact on fish was reported, and everyone contributed something on soil and vegetation. Despite variance in approaches and research, the repeated message was clear: the forest environment of the north was being eroded by acid rain.

When acid-laden rain falls on a typical forest, the raindrop washes over at least three layers of foliage before it reaches the ground. On broad-leafed trees such as maples and oaks the raindrop must pass over much more leaf area than on white pines or other coniferous trees. The raindrop falls on leaf surfaces often dusted with soot, dirt, deposits of dry sulphate, and other airborne chemicals and minerals. It can wash such chemicals off the leaves, or turn them to acids which remain. The sulphuric and other acids can corrode the delicately thin waxy layer which coats and protects the leaves. The acid can directly damage the cells on the surface of the leaf, particularly the "guard cells" which control the opening and closing of tiny pores to allow moisture and gasses to breathe in and out. Damaged guard cells can literally cause the leaf material to suffocate, or be flooded with acid, other chemicals and bacterial diseases. The leaf's normal metabolism can be disrupted; essential photosynthesis can be altered; the abnormal growth or premature death of leaf cells can result. On flowering shoots and buds, acid can contaminate or destroy pollen, interfere with fertilization, even stunt or kill the growth of seeds, or render them sterile before they ever fall to the ground.

This kind of acid-rain leaf etching had been duplicated in laboratory conditions in North America and Europe by the end of 1978.[5] Some of the "laboratories" were actual forests deliberately selected for their remoteness. In the foothills of the Cascade Mountains of Washington State in 1974 and 1975, for example, a group of scientists following up the 1972 Swedish warnings measured acid damage to the needles of Douglas Fir trees, downwind from urban air pollution sources.[6] Acid rainfall there averaged only pH 4.7 to 5.1—cleaner than in eastern North America, but acidic enough to disrupt essential processes. The "scuzz", the almost invisible community of algae, fungi, bacteria and micro-lichens which live on the surface of the pine needles and "fix" nitrogen into a form usable by living cells, had been damaged. Nitrogen is so essential to vegetation that it is used as commercial fertilizer throughout the world and in forests nitrogen-fixing first enriches the leaf, and then the soil. But this wasn't happening in the Douglas Firs of the Cascade Mountains. Trees weren't visibly affected, in the limited time of the Washington study, but the process which keeps trees healthy over decades clearly was.

Time is the critical factor: the length of time an acid-laden rain drop clings to a leaf, the number of times it happens in a single rainstorm, the number of rainstorms, the time of day and time of year, even the particular time in the leaf or tree's life-span when an intense (or slight but repeated) acid fall takes place—all of these conditions influence how soon splotched leaves, stunted needles, retarded fibre growth and thwarted fertilization and germination will result. It has taken more than a decade for scientists to confirm that these effects happen to widely varying degrees among different species in the forests. The US Environmental Protection Agency in 1978 found, for example, that germination of eastern white pine seedlings actually increased to 98 per cent (from 81 per cent) when rain acidity increased from pH 5.7 to pH 4. But sumach and sugar maple, with only a 20 per cent survival rate at 5.7, did even worse at higher acid levels. Such trees are common in the Hubbard Brook forest, now suffering a sharp decline in growth. Such trees were also once common, and now virtually extinct, in the Sudbury area which has endured intense acid rain and sulphur dioxide for more than 40 years. As the first joint Canada-US research report on acid rain, released in mid October 1979, put it: "Effects are cumulative and elusive but to wait long enough to obtain, say, a clearly demonstrated effect of some 15-20 per cent loss in forest productivity, could mean that a stage of forest degradation has been reached that would be impossible to reverse. Several features of the regional situation suggests such a threat exists now."[7]

After acid rain has washed over or burned its way into the leaves of trees, it ends up in the soil, an even more complex and more vulnerable eco-system than the forest canopy itself. A healthy soil breathes—exuding carbon dioxide produced by organisms, bacterial activity and decay; carbon dioxide which can be used as a quick indicator of soil health. Not surprisingly, the Swedish Royal Academy of Forestry was one of the first research agencies to confirm that abnormally acidified soil has abnormally low carbon dioxide respiration.[8] The organisms and bacteria are the key mechanisms to convert dead and decaying matter such as leaves and twigs into the nutrient material which fosters the renewed growth of plants. This rich material, or humus, also binds larger particles of soil together, locks some minerals into fixed chemical formulations, and unlocks others for absorption into the roots of plants. The humus contains nitrogen, the essential soil fertilizer.

Not all soils are alike, although all eventually move to a more acidified state after centuries of slightly acidic natural rainfall which slowly overcomes the natural buffering capacity of the soil and erodes away the humus, the nitrogen and essential minerals. But as scientists have now confirmed,[9] unnatural acid rain accelerates the process. Stripped of humus material and with sharply reduced plant roots to bind the soil, the land around Sudbury, Ontario, has simply eroded away after a mere 40 years of acid drenching. As Hutchison and Whitby discovered, what soil remains is heavily laced with metals such as zinc, cadmium and copper left behind in toxic concentrations when the essential chemical bonds were broken and buffering materials weakened. On deeper, less vulnerable soils, the process may take decades or even centuries to reach a critical stage, but it's only a matter of time. In the midwestern US, scientists have calculated that an annual 40 inches of rainfall at pH 4 (now typical in some regions) could acidify the soil 19 per cent in a century.

The process can be deceptive—in early stages acid can unlock extra nitrogen in the soil, causing apparent increases in soil fertility. But as the acid rain continues, this too is overcome, and an accelerated slide towards prematurely old, unhealthy and unstable soils resumes. The rapid increase in concentrations of aluminum holds particular threats. The world's most plentiful metal, aluminum remains locked in inert formulations under normal conditions. With acidification, it becomes freely-available and this toxic element is easily absorbed into plant and tree roots and is eventually washed into lakes and rivers in concentrations deadly for fish. As on the surface of leaves, acid rain can also boost the growth of harmful bacterial in soil and worse, replace beneficial bacteria with harmful fungi. The worst levels of acid rain are now falling on soils especially weak in buffering protection and high in metal content. Together Canada and the US rain from 15 to 32 pounds of sulphur a year onto forest soils called podzols, the thin covering earth of the Canadian Shield and its boreal forest. Further south, in the Muskokas, Appalachians, Adirondacks, in the White Mountains, in Minnesota and on the Pacific west coast, the acidification process may be slower, by varying degrees, but it is relentless. As more than one scientist has warned, change in soil is the slowest process to detect, but it is the most difficult to reverse or reclaim if it has gone very far. However, if one wants faster, more obvious indications of acid devastation on the terrestial environment, agricultural crops are providing the warnings with increasing consistency.

At the US Environmental Protection Agency's experimental farm in Corvallis, Oregon, scientists have been watering their crops with acid rain since 1977, and watching them suffer. The experiments are meticulously controlled, using soils with concentrations of minerals, metals, acidity, humus, and litter carefully measured and distributed. Acid rain with the same sulphur and nitrogen content as the rain falling on Hubbard Brook Forest, New Hampshire, is sprayed over selected plants and seedlings in greenhouses replicating typical humidity, sunlight and temperature. In some cases frogs and salamanders are added to the mini-environments, and insects, bacteria and fungi are allowed to flourish as they would in the natural state. As far back as 1977 the EPA men with their acid watering cans, spray hoses, clocks and microscopes have been proving the anticipated: acid rain damages crops. With doses of acid rain below the pH 3 level (lower than the present eastern North American average) the scientists could produce pitted leaves on bush bean plants within a few hours. These turned to larger blotches and leaf surface erosion with 24 hours.

The same phenomenon has been discovered on kidney bean, pinto bean, beet, sugar maple and yellow birch leaves. The rain was not enough to destroy the leaves outright. But when the scientists dug up the plants and dissected, dried, weighed and analyzed them under microscopes, they found hidden damages. Even at pH 4—the average level of eastern acid rain—the leaves themselves were abnormally light in weight, their reproductive pods short on vital mineral and nutrient contents, and photosynthetic chlorophyll had been reduced.[10] Across the continent at the Oak Ridge National

Laboratory in Tennessee researchers in 1977 found that strong acid rain reduced kidney bean plants' resistance to certain parasite organisms. They also found that the vital nitrogen-fixing capabilities of the plants' roots was retarded. And in Hawaii where acid rain is caused by sulphur dioxide spouting from a nearby volcano, scientists found they could not grow saleable tomatoes if they didn't keep them out of the rain under plastic shelters.

These were early experiments, under carefully simulated and abnormal or contrived conditions. They were also, unfortunately, almost the only experiments attempted, in the mid 1970s. But by 1979 the EPA's Corvallis labs had extended their crop variety, and had moved out of the greenhouses into the fields. The Corvallis labs are now testing the effects of simulated acid rain on at least 30 different plants representing every major field crop grown on the continent. In September, 1979, in Washington, EPA acid rain chief Dennis Tirpak showed the authors the results of one of the first crop test results: ordinary radishes suffered a steep drop in the fresh weight of their roots as soon as regular acid rainfall dropped below pH 4. "That's only the beginning. We're hoping to show exactly what this means in actual crop weight, growing period, and susceptibility to other environmental factors for more major crops, within another year," Tirpak said. "We can't be sure about what we'll find, but I'm sure we've ignored this aspect of acid rain for far too long."

As a Canadian scientist put it in another interview, "we're barely managing our ignorance about the effects of acid rain on crops; we are far from having any scientific conclusions. But what we do know is not at all encouraging." What is known is that radishes, tomatoes, beans, lettuce and beets are affected by acid rain stronger than pH 4. The effects so far have been reduced weight, inhibited nitrogen-fixing, and leaf damage, but not destruction. The damages were largely produced in artificial environments—greenhouses, carefully guarded test fields—and over brief periods of time—one season, usually. But do the experiments duplicate the real environment? Scientists do know that equally strong acid—pH 3.9—occurs at the very height of the growing season in rich vegetable-growing areas of Pennsylvania. They know that experimental tobacco crops suffer up to 40 per cent leaf damage in summers when the flow of ozone associated with nitric acid rain is particularly heavy in the Stouffville area just north of Toronto. They know that apples develop unexpected scars and blotches in the area of New York state where acid rainfall is heaviest in summer. But they don't know what other factors could be at work—soil acidity, microbacterial activity, other airborne pollutants—in thousands of other prime agriculture areas across the continent where it rains acid. Scientific ignorance of the effects of acid rain on agriculture is presently as immense as the area over which acid rain is falling.

Certainly in some areas of intensive agriculture, soils are sufficiently buffered with neutralizing chemicals and can be, and are, regularly fertilized with neutralizing elements, to stave off any sudden acidification. Southern Ontario may be one such fortunate spot, although there are areas of abnormal soil acidity in the prosperous Essex, Middlesex and Kent County region. But will the anti-acid capacity of the soil be sufficient to counteract the acid stress directly on the leaves, stems and seed pods of crops? For how long? And what of the thousands of square miles of agricultural land which is not fertilized, or the even larger areas of untended grazing land? What effect has acid rain had, what effect will it reveal, and how soon? At present, there simply isn't enough scientific information to answer these questions.

And yet, there are those who use this lack of knowledge to imply that there is no problem. Scientists from the Ontario Environment Ministry, for example, assured the Legislature committee investigating acid rain in early 1979 that "at present there is no direct proof that ambient acid rain is having any significance on native vegetation." The statement was true, as far as it went, but it failed to emphasize how little is known. One of those same ministry experts, Dr. Sam Linzon (who had documented sulphur dioxide fumes affecting trees near the

Sudbury smelters in 1950) offered similar assurances later in 1979. During a Toronto briefing of environmental groups in October Linzon pointed to a 1976 Swedish report on effects of acid rain on forest vegetation and soil. Linzon had even under-lined what for him was a key sentence in the Introduction: "Consequently there is no direct proof that rain of the acidity commonly occurring in southern Scandinavia has adversely affected tree growth." The report, by Carl Olaf Tamm, noted the difficulty of linking acid rain to immediate and measurable impacts on wilderness forests but went on to document results from laboratory and experimental field studies. However, what Ontario government scientist Linzon had not underlined was the equally important *conclusion* in the Tamm report: "Consid-ering that some soil processes *have* been affected in acidifica-tion experiments, the most likely conclusion seems to be that Scandinavian forest ecosystems *are* affected by acid rain in a direction which means less productivity in the long run." (Emphasis added.) Nor did Linzon choose to point to a similar conclusion of the first Canada-US scientific assessment of acid rain, released days earlier, which warned "there is every indica-tion that acid rainfall is deleterious to crops."

As the Ontario legislature committee commented in its 1979 report, absence of definitive proof of acid rain damage is no basis for concluding that such damage is not occurring. The risks of elusive cumulative and even synergistic impacts must be recognized. The committee report also noted there is every indication the public has not been alerted to these risks. There is a truism which says ignorance is bliss. But to paraphrase a more important dictum, ignorance is no defense in the eyes of the law, including the law of nature.

Despite the evidence, few scientists are willing to hazard more than suggestions for "further research" into narrowly-defined areas such as specific soils or vegetation. And yet beyond these narrow visions lies a wholly inter-linked eco-system which fosters life itself on this small planet. Already there is evidence that the biological food-chain is weakening, with ominous consequences for higher mammals. Fish and aquatic organisms die off in acid lakes. With fewer organisms and insects able to breed and feed, other wildlife suffers a decline in food supply. Amphibians such as frogs, toads and salamanders are next affected, and they already have enough problems withstanding the direct shock of acid waters. In the Adirondacks, for example, researchers have found that spotted salamanders' eggs are quickly deformed and fail to hatch when anything more acidic than normal rainfall (pH 5.6) accumulates in ponds. The salamander population in the Adirondacks is abnormally low today, as is the pH of the rain and snow which falls there.[11] In the acidified Killarney Lakes, the frog population appears to have declined, and in Scandinavia the same situation has already been confirmed.

The salamanders and frogs used to feed on insects; raccoons, skunks, birds, shrews and foxes used to feed on the amphibians; larger mammals in turn feed on these—the effects of acid rain are moving up the biological pecking order. In Norway the population of capercaillie and black grouse has declined, possibly due to direct acid poisoning of insects or shortages of these vital foods for young grouse. On the west coast of Sweden fish-eating birds such as mergansers have migrated away from acidified lakes in search of better supplies of food. And in central Ontario the loon, a bird whose haunting cry has long symbolized the clean and natural environment a million cottagers imagine they are enjoying as they sit on their docks and shorelines at sunset, is heard less and less often. Certainly further research is needed. Of course, many factors contribute to the decline of animal populations. But while passing on reports and passing the buck, we might take time to remember that if we're going to poison the very basis of nature's life cycle, the implications reach to the very top.

## Illustrations

*page 71:* Richard Beamish.

*pages 72–73:* Lars Overrein.

*pages 74–78 (top):* Charts adapted from the US EPA *Research Summary: Acid Rain*, October 1979.

*page 78 (bottom):* Adapted from the report of the International Joint Commission, Great Lakes Science Advisory Board, July 1979.

*page 79:* Based on data from the Ontario Ministry of the Environment.

*page 80:* Adapted from Frantisak and Hunt, *Long Range Transport of Acid Compounds*, Pollution Control Association of Canada, April 22, 1979.

# Deformed Adult Fish from Acidified Lake, Killarney, Ontario

# Fish Deformed by Acid Waters

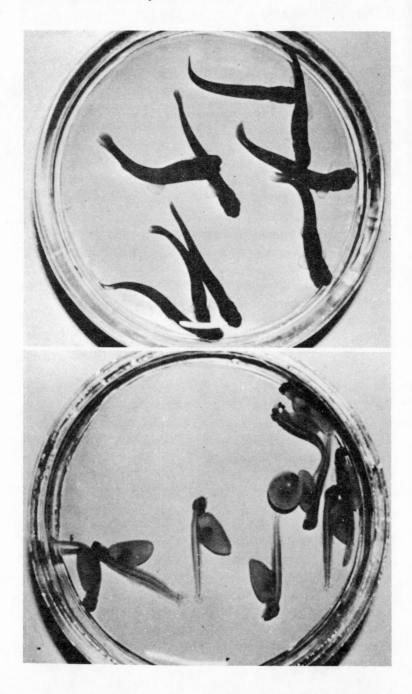

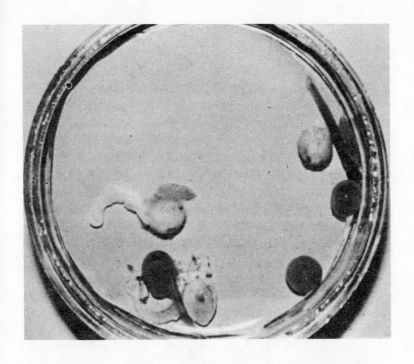

Above left: In "clean" rainwater (pH 5.5) trout fingerlings hatch normally.

Lower left: In moderately acidic water (pH 5.0) fingerlings' growth is distorted.

Above: In water as acidic as many Ontario lakes (pH 4.6) fingerlings are unrecognizable.

# Acid Rain in North America: Then and Now

Annual Average 1955-1956

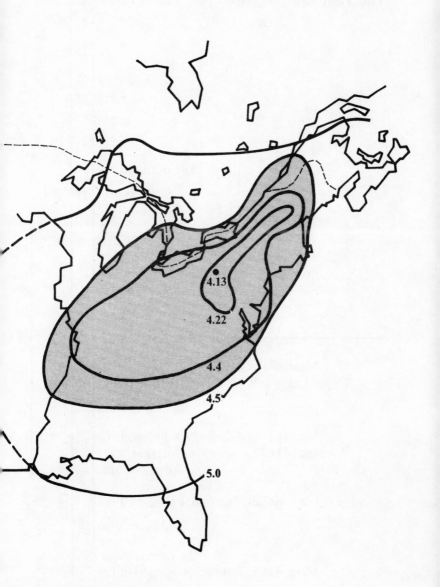

Annual Average 1975-1976

# The Acid Scale

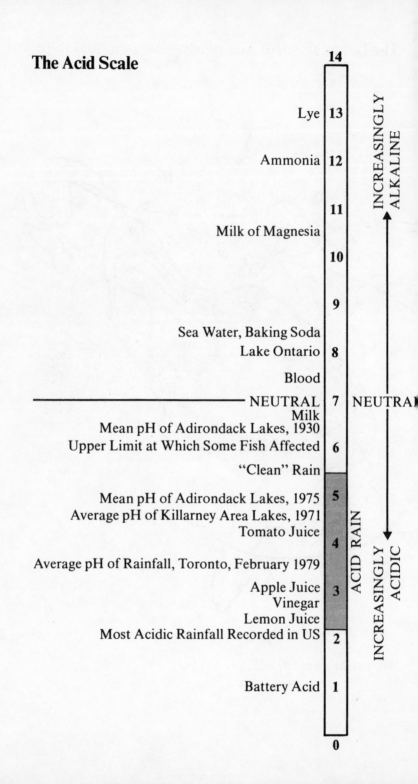

# The Lakes of North America Sensitive to Acid Rain

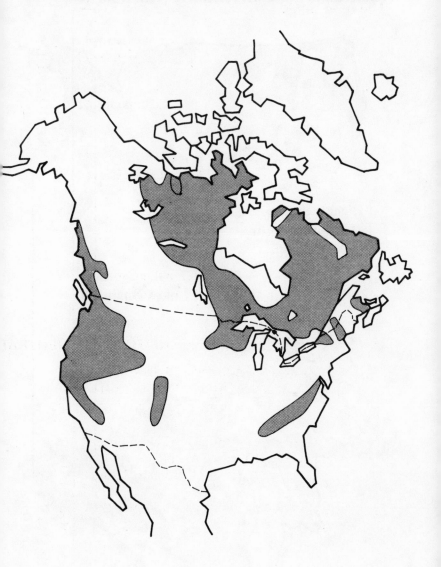

# Vulnerability of U.S. terrain to Acid Rain

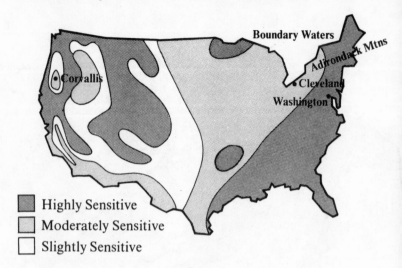

Boundary Waters

Adirondack Mtns

Corvallis

Cleveland

Washington

Highly Sensitive
Moderately Sensitive
Slightly Sensitive

# Acid Rain Now in the Great Lakes Basin

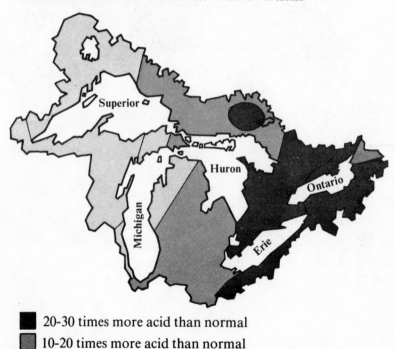

Superior

Huron

Ontario

Michigan

Erie

20-30 times more acid than normal
10-20 times more acid than normal
5-10 times more acid than normal

# Vulnerability of Ontario Terrain to Acid Rain

■ Highly Sensitive
▨ Moderately Sensitive
□ Slightly Sensitive

# Acid Rain in Canada, Midsummer 1977

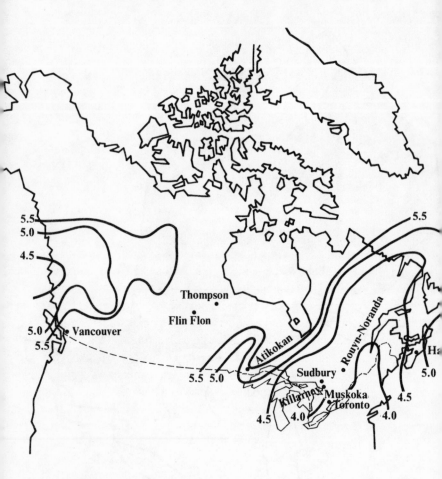

# 4
# Acid in the Community

So far, research on acid-rain damage to the terrestrial environment remains almost unnoticeable compared to the size of the problem, and is couched in cautious technical papers and reports. But there is one impact which is obvious to many North Americans: the decline of visibility. As the US EPA's acid rain study co-ordinator Dennis Tirpak outlined in an interview in September 1979, only the New Jersey-Connecticut region of the eastern US had an average visibility of less than eight miles, back in 1958. But as reports from airport and other weather observers showed, the low-visibility area had doubled by 1968. And by 1973 almost *all* of the United States east of the Mississippi suffered a maximum visibility of eight miles. Three regions between New York and Richmond, Virginia, were cut to no better than four miles. Tirpak linked that sharp reduction in visibility to sulphur dioxide (and carbon monoxide) pollution which had sharply increased over the same period of time. The kind of acid-ozone smog which had once been a rarity in most cities, notably excepting Los Angeles, has now become widespread on the eastern seaboard downwind from the smokestacks of the Ohio-Mississippi Valley power plants and heavy industries. Similar increasingly frequent haze has been charted over Atlantic Canada, where much US air pollution ends up. It is no wonder that aircraft landing in New York are often forced to rely on radar and instrument guidance all the way to the ground, during the daytime.

The growing acidic haze is more than just a nuisance to

aircraft navigation. It reduces the passage of sunlight to the earth, which in turn could upset oxygen-generating photosynthesis in plants, the processes of evaporation and condensation which move vital moisture to and from the soil, vegetation and the atmosphere, and the warming and cooling of the land and low-level air masses. It is too soon to tell how much of an effect pollution haze has already had. It could produce a warming of average temperatures as warm air is trapped below the haze, or a cooling as less sunlight penetrates the haze. Scientists aren't sure yet, although they have warned that as little as a 2 degrees Celsius temperature change could have devastating effects on regional weather and agriculture. The same scientists have already sounded loud alarms about a related atmospheric problem—the build-up of concentrations of carbon dioxide around the globe at a higher altitude, which is creating a "greenhouse effect". Like acid haze but with even greater effectiveness, the carbon dioxide is reflecting more sunlight and heat back into space, and possibly trapping more warm air close to the earth. An ultimate warming or cooling effect is still unclear. What is known is that carbon dioxide levels have been rising 3 per cent per decade for more than 50 years and by 1999 will have doubled from pre-industrial levels. And by the beginning of the 21st century the world may be facing "social, economic, environmental and economic consequences" bigger than anything previously imagined.[1] In 1977 the US scientific community declared carbon dioxide build-up and its consequences to be one of the greatest environmental threats facing the continent, and the whole world. The carbon dioxide comes from the same sources as acid rain—the smokestacks and exhaust pipes of the heavily industrialized, mobile industrial society. In 1978, the US presidential Council on Environmental Quality added acid rain as an equally critical threat, at least as far as North America is concerned.

But declining visibility is not the only impact of acidic air pollution on the human environment. As far back as the mid 1940s the Swedes had studied air pollution corrosion rates of typical metals.[2] Between the mid 1940s and mid 1960s, the rate of corrosion of the test metals doubled in rural areas, and increased six times in urban areas. The increase was blamed on the larger volume of sulphur dioxide and dirt in the air. Galvanized steel plate, for instance, would last only 4 years in urban air before it needed painting to cover pit marks and surface corrosion. The same type of metal in relatively clean air in rural communities far north of the urban and wind-blown sulphur dioxide could last 25 years unmarked. Between the 1940s and 1960s, the flow of acid-rain laden air over Scandinavia had increased sharply.

Even steel painted with traditional anti-rust paint was not immune to the sulphur dioxide which penetrates and weakens paint's adhesive abilities. Painted wood had been equally vulnerable, the Swedes warned, until the mid 1960s' introduction of less acid-susceptible chemicals for paint production, particularly the all-synthetic latex compounds. But for materials which are rarely painted, such as sandstone, limestone, concrete and plaster, the Swedes documented serious deterioration. The calcium in these materials was turned into crumbling calcium sulphate by the presence of sulphur dioxide. In some cases the lime used in cement ended up being leached out in white salt-like stains.

Since that time it has been confirmed that the principal corrosive component of air pollution which has eroded the Parthenon in Greece, the pillars of Venice, the Colosseum, and even some monuments in North America, is sulphur dioxide. In most cases the air pollution is localized, but not always: now US investigators are checking tombstones across the continent as possible signposts for the rate of acid rain damage. And in 1978 the US presidential Council on Environmental Quality offered a rough estimate that acid rain was costing about $2 billion annually in architectural and statuary damage. Too little work had been done to predict how soon serious structural damage might develop.

Smoggy skies, blistered paint and corroded metals are visible to all of us but a much less recognized problem is what acid rain may do to the water we drink. In 30 communities studied in the Adirondack Mountains of New York state, health department officials found that public water facilities were abnormally corrosive—not necessarily strong in acidity but very weak in buffering capacity—so weak, in fact, that the addition of chlorine only made the water more corrosive. And as New York state health specialist Dr. Wolfgang Fuhs discovered in some residences, copper piping was being corroded by this water.[3] Hot water is more corrosive than cold water, and when the water is allowed to stand for several days, copper levels can rise above the maximum US drinking water standard to a toxic concentration. The same corrosive water leaches lead from solder joints in metal pipes, and Fuhs' researchers found excessive levels of lead in private well water even before it was drawn into the pipes. (Lead is a subtle, cumulative poison, particularly dangerous to children.) These discoveries were made in an area suffering some of the worst acid rainfall on the continent, admittedly one with old water systems and private wells which lacked adequate treatment systems. To the end of 1979 no cases of metal poisoning had been discovered, but the state has now advised the area residents to flush their pipes once a day before use and not to use copper pipe in new construction. In Sweden where the same problem exists, some municipalities are required to add neutralizing chemicals to their municipal water supply. New York state has also discovered that outbreaks of gastroenteritis in the Adirondacks are occurring with the rise of a normally rare but acid-resistant bacteria in the acid-laden water supplies. And as well-established studies in Atlantic Canada and Newfoundland have shown, there is a greater incidence of cardiovascular disease in areas supplied by softwater (low alkalinity) acid-buffeted waters.

In Ontario the environment ministry considered the risk of acid corrosion of water pipes, and rejected it, although Ontario's waters are strikingly similar to those of the Adirondacks. In April 1979 Ontario environment minister Harry Parrott, replying to a Toronto *Star* article citing US reports of corrosion by acid water of lead from old water systems, had written that there was no need to worry in Ontario. Most community and private water pipes are made of copper, not lead, he said. In June, when Opposition Leader Stuart Smith correctly pointed out to Parrott that the New York reports involved copper pipes and lead solder, Parrott went so far as to agree that "perhaps on further reflection it could be safely assumed that [his] statement [to the Toronto *Star*] isn't a total reflection of the fact."[4]

Parrott's rambling reply to Opposition Leader Smith included the argument that cottagers on the acidified lakes of Ontario abandoned their summer homes in the fall after draining their water systems to avoid frozen and cracked pipes in the winter. He did not discuss short-term summer supplies of water. He did not discuss the need to check communities using acidified lakewater as permanent supplies. And he said it would be "an almost impossible job to notify all the cottagers ... It would be perhaps impossible." Parrott did admit that the warning to cottagers to avoid water which had stood in pipes too long "is fair and people should use that caution." He added that the fact that the issue had been raised "perhaps might be sufficient notice." But most Ontario residents do not read the transcripts of Ontario Legislature debates.

Given such leadership and resolve on such an issue as water pipes, it's scarcely surprising that the larger effects of acid rain on human health have been so ignored. The Ontario Legislature committee report did mention that there was a "suspicion" that "more negative effects relevant to human health will occur and be uncovered in the coming years." The joint Canada-US report in 1979 did not mention health at all; what government wants to talk about such possibilities? Yet the direct effects of sulphur dioxide at high concentrations have been well established for years: sore throats, coughing, lung irritation, even serious lung tissue damage. Both Canada and the US have limits on permissible concentrations in ambient or "typical" air. When these levels have been exceeded, the incidence of respiratory ailments has noticeably risen. In extreme and prolonged sulphur dioxide pollution such horrors arise as Britain's "killer smog" in December 1952, when more than 4,000 died. (It was the risk of repeated episodes which prompted Britain to curb urban coal burning and to build tall smokestacks as the key to cleaner British air—much to Scandinavia's later sorrow.) Such episodes are rare in North America and the Canadian government in particular has repeatedly argued there's no significant health risk from sulphur dioxide at typical concentrations.

But what of sulphur dioxide which has turned to sulphates, in the form of acid or as dry particles? The Swedish government in its 1972 report to the UN warned there was mounting evidence that "small droplets or particles containing sulphate or sulphuric acid are capable of penetrating much further ... into the narrowest passages of the lungs." This could happen even when sulphur dioxide is at apparently "safe" levels, and is absorbed and neutralized in the nose and throat before it reaches the lungs. The Swedes expressed concern about the possibility of acid rain harming human health, but admitted a lack of studies to confirm their suspicions. Seven years later, as the Ontario Legislature committee noted in early 1979, "there is an absence of scientific certainty about the extent and severity of the human health effects of sulphur dioxide, sulphate particulates and acid rain singly or in combination." A few months later Dr. Leonard Hamilton, head of environmental assessment at the Brookhaven National Laboratory in New York, suggested that the components of acid rain might be killing 5,000 Canadians per year.

As Dr. Hamilton explained at a Toronto acid rain conference, the Canadian figure was simply an extrapolation from American figures of at least 50,000 deaths per year. Canada's acid rain patterns, wind directions and relevant population distribution was similar to the US acid rain-impacted areas, and the actual process was exactly identical—tiny particles of sulphate are carried into the lungs, damaging the tissue and causing bronchitis, emphysema and a strain on the heart and circulatory system; basically what the Swedes had suspected seven years earlier. As Hamilton emphasized, this mortality rate was low compared to other hazards: only 2 per cent of total US annual deaths compared to 12 per cent from smoking, for example. But the relationship between acid rain and health was clear, Hamilton said, citing examples of increased lung ailments reported after every major 24-hour incidence of elevated sulphates in air over New York city.

With 10 to 20 micrograms of sulphate in every cubic metre of air in the eastern US on a regular basis, there are far more than enough particles of sulphate floating freely to be absorbed into the lungs of thousands of people already susceptible to lung and heart ailments, and in many cases ostensibly healthy people, to cause at least 2.7 per cent of all the deaths recorded in any one year. Hamilton's conclusions gained little media attention at first—they were the first strongly asserted indications of direct acid rain-health links, and early bearers of large scale environmental crises are accorded much less space in print these days than those who claim wonder cures and medical breakthroughs. Harold Harvey faced the same problem in 1971 when he warned that Ontario lakes were dying.

More evidence is surfacing now. Robert Mendelsohn of the University of Washington and Guy Orcutt of Yale came up with a figure of 187,686 deaths per year in the US from sulphate in the atmosphere in a report published in June, 1979.[5] They came to their conclusion after combing more than 2 million death certificates, the census data for 3,000 counties, and detailed sulphate air pollution reports from annual US Air Quality Reports by the Environmental Protection Agency. By means of computerized assessment and calculation, the two academics found that increased levels of sulphate in air matched increased deaths due to all causes but particularly to heart disease. Their findings "present striking evidence that sulphates are deleterious to health ... a definite association between pollution and deaths from heart and circulatory failure."

They also noted that "there is no evidence of a lower threshold"—in other words, there is presently no known minimum concentration of acid rain-borne sulphates which will not increase a large population's death rate. "If society wants to eliminate air pollution effects, no one can be exposed to the lowest existing level of...the harmful air pollutants." And since the worst air quality in the US is in the north-central and northeastern regions, "the probability of dying from air pollution in these two regions is about twice as high as that in the rest of the country." Canada has vastly larger areas of relatively clean air, but very few Canadians live in those areas. Two-thirds of the country's population live clustered close to the border, adjacent to those American regions of most-polluted air. There is good reason to suspect the same forbidding assessment of air pollution mortality could apply to southeastern Canada. After all, much of the winds blow from the US to Canada.

The Mendelsohn-Orcutt study has not been previously reported to Canadians in any public media. It was first published in an obscure scientific journal, not the sort of thing a layman normally sees. More studies may already exist in equally obscure journals, and some may even contain important criticisms and refutations. Harold Harvey's first reports of dying lakes in the Killarney area came out in similar journals, and it took five years for them to reach a larger public. How long will it take before the health dangers of acid rain are brought to the larger public's attention? The journals, reports, and conferences are supposed to be part of an early-warning system and peer-review mechanism which keep the scientific world informed. Scientists in turn are supposed to bring the products of their research to the decision makers. Two decades of billion-dollar expenditures on reviving the poisoned Great Lakes, driving reluctant auto-makers to the pollution-control wall, and evacuating whole neighborhoods from chemical sewers like the Love Canal show how important our environmental resources are. As the monumental cost, human suffering, and unsettling political and economic revelations have proven, reaction to pollution is usually far too little, too late; prevention is the only cure. And yet, preventative acid-rain information from the scientific community seems to be scarce in North America. Never has a contaminant better suited the Barry Commoner maxim: "everything affects everything else." Poisoned skies mean dead lakes mean poisoned land mean poisoned vegetation mean poisoned food and air. And yet scientific research on acid rain has sounded the warnings only when each step in the eco-system is already proven damaged or dead. Where are the early warnings of what needs protection? The guardians of the public's vital interests have become jealous wardens of imprisoned information.

An incident at the Bedford Institute of Oceanography shows why North America has only begun its education in acid rain. The Bedford Institute of Oceanography lies clustered along the shoreline directly across the harbor from Halifax, its buildings perched on the bare rock outcroppings which typify much of the southern half of Nova Scotia. In mid October 1979, 85 North American scientists met at the Institute for a two-day assessment of what acid rain has done to easternmost North America and what to do about it.

The meeting was held in a boardroom near the huge John A. Macdonald bridge which funnels tens of thousands of vehicles across the harbor and paralyzes the traffic in a twice-daily jam-up. Beyond the windows of the boardroom three giant candy-striped smokestacks of the local power plant added hazy contrails to Halifax's notorious foggy grey skies. As Dr. Rod Shaw, chief of Environment Canada's regional air pollution division commented in the opening minutes, the meeting was long overdue. Atlantic Canada lies like the narrow end of a funnel of prevailing weather patterns which rise over Ohio, New York and Ontario and converge on the east coast. Twenty out of every 30 days in a month, the wind blows down the funnel bringing the sulphur and nitrogen oxides, the heavy metals and airborne grit of industrialized North America to the east.

Early sessions of the meeting were spent reviewing what was already well-known: Atlantic Canada was suffering intense acid rainfall. More than 35,000 tons of sulphuric and nitric acid annually rain on Nova Scotia alone; 600,000 tons of sulphate was falling on the Atlantic region; 7 to 15 pounds per acre. Regional rain had intensified 10 times in acidity in 20 years; the number of hazy, poor-visibility days had tripled. And inevitably the rain had done its damage. The salmon population in major rivers like the Moira had been cut in half because the waters had dropped to pH 4.6 by 1978—far too acidic for newborn salmon to survive. The thin soils and bedrock of Nova Scotia lacked depth and chemicals to neutralize the acid. Other fish were declining in the spring acid shocks. As Walton Watt, a federal fisheries expert said in an interview, "the evidence of fisheries damage alone is enough to convince me we've got to act immediately to turn off the acid rain."

Watt's colleagues from Canadian forestry research institutes were much more restrained. Dr. Peter Rennie, an environmental policy analyst, spent much of an afternoon reviewing different theories of possible acid rain damage to forests and vegetation, some of which point to inevitable damage over time. But he concluded there was no proof of any theory's validity. Dr. Keith Puckett, of the federal Atmospheric Environment Service, was more direct in his presentation: "there is nothing to suggest any direct or indirect effect of acid rain on vegetation." And Dr. Surin Sidhu of the federal Forestry Research Centre in Newfoundland reported that "nothing much has been achieved in the past 2-3 years" in charting vegetation susceptibility to acid rain.

The three scientists' conclusions were startling in their staunch discrediting of forest and vegetation damage and their apparent limited concern. Taken at face value, the statements contradicted the joint Canada-US science report on acid rain which had been released (after months of delay) only four days earlier. That report, which had been deliberately drawn to the media's attention, had warned "there is every indication that acid rain is deleterious to crops." It had also pointed out that the Canadian forest industry, working under a cloud of acid, was an enormously valuable resource. As more than one worried Ontario scientist had pointed out during the slow rise of public awareness about acid rain in 1979, it would take the threat of enormous financial loss in agriculture and forestry (health was not yet considered) to move the public to demand an end to acid rain. Dead fish and lakes would not be enough.

Yet Rennie, Puckett and Sidhu seemed to be concluding such risks were non-existent or, at best, exaggerated by the media. An interesting unscheduled exchange broke out between the three "experts" and a visiting journalist and an environmental group director on what the public was to make of such assertions of unconcern. Pressed to explain, Rennie replied that "there are no clear, unequivocal documentations of damage to vegetation." Reports of crop damage in Michigan, the US experiments which produced a 50 per cent drop in fresh vegetable weight, and the Swedish studies were offered in reply as "important". But Puckett replied: "Adverse effects on crops are found only in simulated experiments, as far as I know. We have no data based on research in the field." And Rennie added that the reported Michigan bean field crop damage might be due to not only acid rain but also other factors. He insisted he was not prepared to link acid rain to any damage anywhere. He hadn't seen "anything in the field." None of the three experts commented on the Swedish studies, none mentioned Hubbard Brook. And despite pressure, none of the three would say what they expected or predicted to find concerning studies of acid rain and vegetation. Sidhu did allow that "it is possible there will be a 20 per cent reduction in forest growth and we'd still not be able to detect it." But clearly, without definitive proof based on years of close examination, none of the three were prepared to offer hypotheses, predictions or prescriptions for preventative action to avert initially undetected damage.

It was a frustrating exchange, one which stood in stark contrast to the opening statement to the Institute meetings by Environment Canada's Atlantic Region director: "We run the risk of doing too little until the damage is incurred. Can we wait until all the impacts are studied, before starting on acid rain controls? No, clearly we can't." As the speaker, Dr. C.J. Edmonds, politely reminded his colleagues, their profession's traditional conservatism and distaste for aggressive, predictive stances was itself risky. The idea of extreme environmental change over the entire Atlantic region, or much of eastern North America, was "too sobering a thought" to allow scientists the luxury of total and almost abstract detachment from the political realities of acid rain, Edmonds said. Rennie, Puckett and Sidhu chose not to heed his gentle chiding, a position not entirely unexpected from Canadian government scientific agencies, seldom known for outspokenness or political independence. The three experts confined themselves to being terribly scientific, but not very helpful. None of the three chose to comment on how their assessments would figure in any moves to curb acid rain as quickly as possible. The joint Canada-US report—which Rennie had contributed to—almost presciently commented on the Rennie-Puckett-Sidhu approach to the problem: "To await an unequivocal demonstration of damage (for example a 15-20 per cent loss in forest productivity) is clearly unacceptable. By that time economic loss could be great and site degradation would be irreversible."

But the Institute acid rain sessions were "scientific" in nature and much questioning about the real implications of acid rain on forests from the invited media was clearly an unnecessary interruption. The meeting adjourned for coffee. Fortunately there was another interpretation about the relevance and implications of acid-rain forest damage studies. Minnesota biologist Eville Gorham, a former Nova Scotia researcher who had moved on to Sudbury and Scandinavia to become advisor to the US presidential Council on Environmental Quality report on acid rain, was somewhat outspoken on what his fellow scientists had talked about.

In an interview with the authors, Gorham offered the opinion that in the face of an acid rain which presently appears uncontrolled until the end of the century, scientists have a clear social and political responsibility. "In something of this magnitude, I can't see the sense of sitting isolated in some ivory tower reporting what has already happened, if it can be prevented by some reasonably-based prediction and advocacy." He agreed that at the least the existing evidence "logically points toward damage being caused" by acid rain to forests and crops. "That's reason enough to advocate controls on the pollution." He also agreed that the public and corporate distaste for the cost of controls won't be overcome by scientists who refuse to sketch out the likely consequences of no action. "I'm sure it was impossible to convince the Romans that their empire was about to collapse," he observed, "And it took 4,000 deaths in the London killer fog of '52 to change England's mind about pollution. But does it require a catastrophe here to change our minds about acid rain?"

But only very reluctantly would Eville Gorham, biologist, university professor and US presidential advisor on acid rain, make his own prediction: "Yes, I anticipate acid rain damage to forests and crops." And Gorham was thus outspoken only in a private conversation, not in the Institute boardroom full of fellow scientists.

Gorham's reluctant effort at scientific advocacy came more than a year after Dr. Tom Brydges, an Ontario government water toxicity expert, had offered a succinct explanation of why acid rain has been so long neglected: "In hindsight, I guess we lacked the foresight to take it seriously." He was commenting on why North American scientists had ignored the 1971 Swedish warning that acid rain devastation there "might possibly exist in similar situations in North America...and a detailed study of the likelihood is a matter of urgency." The urgency, however, wasn't clear to most North American scientists in 1971, 1973, 1975, or even 1979, at least in Halifax.

That "likelihood" is now becoming proven as an inevitability. But the foresight to warn the public still remains unclear, to at least some scientists.

# 5
# Acid Economics

Fishing-lodge operators in northern Ontario are none too optimistic these days. At more than three-quarters of the 1,600 lodges and road's end resorts across the north the owners say prospects for the next five years are even chancier than usual. Having poured more than $65 million of their own money into buildings, cabins and boats since 1972, they have developed a canny sense about their industry and its resources. As they told government researchers in a 1978 study, their costs were being driven up and their resources, such as the yellow pickerel, northern pike, and brook, rainbow and lake trout which attracted the Americans who make up 65 per cent of the clientele, were declining.[1] Nearly 85 per cent of those Americans who came (on average) 700 miles to the north and spent $675 per visit, came for only one reason: the fishing. And as the northern affairs ministry researchers later reported, those Americans, and Canadians, generated more than $120 million in annual direct and indirect revenue, creating more than 10 per cent of the 200,000 jobs available in the north.

As the government report also noted, the further east across the wilderness of Ontario the researchers moved, the more and more discouraged they found the fishing-lodge operators. In the Sudbury, Timiskaming and Nipissing regions of northeastern Ontario, which support nearly half the northern lodges, one-half to one-third of the operators were bluntly pessimistic about their future prosperity. The report did not mention that the northeast was the area most heavily rained upon by acid, and preferred to attribute fish decline to cyclical populations, over-fishing, and unspecified urban and industrial pollution. The report offered numerous recommendations for more studies. Increased fish restocking was suggested, as well as a public-relations campaign to remind us all that our aquatic resources are under "stress." It also included the ominous warning that "fishermen are going to have to accept declines in quality standards."

But as one fishing-lodge operator discovered in 1979, much more than just "quality" was at stake. Jerry Liddle, a young operator whose family runs three lodges in the northern Wawa area, approached the provincial environment ministry in 1978 for funds to conduct a major study of acid rain and its implications for his industry. As Liddle correctly suspected, nothing like this had been considered in any detail by the provincial government (and certainly not by the northern affairs ministry study). Liddle however received only a few thousand dollars, and not from any of the policy-making, long-range research divisions of the environment ministry but from the already alarmed and over-burdened fishery and water quality team. The funds barely covered Liddle's costs of duplication and postage for a simple questionnaire, but it was enough for him to come to some rough conclusions by late 1979.[2] His fishing lodge colleagues knew little or nothing about acid rain—but they knew their fishery was declining. As he summarized, "the trend is towards smaller and smaller fish, a lack of large or spawning fish, and increasing difficulty in catching fish." His summary bore remarkable similarity to the conclusions University of Toronto researcher Harold Harvey reached exactly a decade earlier concerning the Killarney lakes of near-northern Ontario, the lakes which had been killed by acid rain.

"The quality of the resource base has been going downhill, especially in the last 10 years," Liddle added. "How long can the fishery stand this pressure and how long will the industry be able to survive?" Based on environment ministry data on the beleaguered lakes of northeastern Ontario, Liddle came up with an answer: almost 600 fishing lodges could go belly up within 20 years if the acid rainfall continued, killing about 6,000 jobs and $28 million of annual income in the area. The environment ministry received his report with little comment, but Jerry Liddle had uncovered the tip of an iceberg. The fishing lodges Liddle studied serve only 12 per cent of the more than 16 million fishing "occasions" which take place in an average year. According to government statistics almost one out of every two Ontario male residents goes fishing, and one out of three females, spending an average of $154 per year on their hooks, rods, bait, hipwaders and so forth. Combined with the Americans who come north, these fishermen spent an incredible $450 million in 1975, a typical year. The average Ontario fisherman travels less than 350 miles, which means using the already over-fished near-northern areas of Muskoka-Parry Sound, Haliburton, and Sudbury-North Bay. As both Jerry Liddle and provincial studies show, there are already five people fishing for every available fish in the northeast. And this is where acid rain is falling most heavily, on the 140 lakes already known to be acid dead and on the estimated 48,000 more lakes similarly jeopardized.

To mid 1980, there were no Ontario government studies on the financial impact of dead lakes. In fact, it wasn't until early 1980 that the first such study was commissioned, and it could take two years or more to complete. Fortunately, to the south in the 6 million acre Adirondack state park in New York, where equally intense acid rain is monitored, there are indications of what Ontario may find. The park lies only a full day's drive away from more than 55 million Americans, and each year, until recently, at least 1.7 million fishing trips are registered in the park, generating an estimated $16 million in the local economy. But in 1976, after the confirmation of more than 100 acid-dead lakes, state park researchers estimated that nearly $1.5 million in fishing expenditures had been lost. As parks commissioner Anna La Bastille told a Toronto conference on acid rain in late 1979, fishermen don't spend money to dabble in dead lakes. A more detailed study in 1978 based just on the dead lakes—the total had risen to 170—showed a direct annual loss of $370,000. Applied to all 3,000 lakes in the park, the economic loss was estimated at "probably much higher than $1.7 million per year," the commissioner reported. "We've turned that park into a national acid cesspool, and now we in the area are paying for it," she added.

But the Adirondack figures are only a foretaste of what Ontario soon could be paying for the acid in its 48,000 threatened lakes. The once-clean water resources of the near-north attract millions of visits, from residents and tourists alike, for the fishing, recreation and cottaging. In Ontario as a whole, tourism is the second largest industry in the province, directly accounting for $5 billion in annual revenue and 470,000 jobs— nearly 6 per cent of the total Gross Provincial Product and 11 per cent of all the jobs.[3] The southern cities draw the largest amount, but nearly $900 million is spent just in the area from the Bruce Peninsula to the Muskoka-Haliburton boundary and all of the northeast. Much of that area is the most acid-vulnerable too. It's the cottage heartland of Ontario, containing almost two-thirds of 250,000 such get-away places which foster 50 million person-days of relaxation each year (the equivalent of 7 million people each spending one week "at the cottage"). Those cottagers directly spend $200 million per year in the area. In an optimistic tourist-promotion study in 1977, the provincial tourism ministry noted that Ontario "possesses impressive water resources, difficult to match anywhere in the world" as tourist attractions, and ranked the 17 most important tourism areas of the province. The fourth most important area was the Barrie-Parry Sound-Huntsville triangle. Ninety per cent of the visitors in that playground come from Ontario itself, and with outsiders, spend $96 million per year for a holiday. But what the tourism report failed to note is that the Parry Sound-Huntsville area is dead centre in Ontario's acid rain zone, where acid rain falls at levels difficult to match anywhere else in the world.

It's not easy to quantify what a lake with no fish means to a cottager with a $40,000 investment in his summer chalet, water-ski boat, lounge chairs and beer cooler. Or whether the lack of fish, frogs, lake-feeding animals and birds means anything at all to city dwellers who drive for hours each weekend just to be able to sit beside the increasingly clear waters. Ontario environment ministry officials claim that residents of some acid lakes near Sudbury actually like the crystalline water (although there's nothing alive to see in it). But for 70 years city dwellers have been spending massive amounts of money for the privilege of owning a piece of nature by an unspoiled lake. The Ontario government uses woodland green for its official lettering, fitting for a province which fosters more wilderness summer camps than anywhere else in North America. The most popular "tourist" activities of Ontario residents are cottaging, boating, fishing and camping. The premier of the province may tour the world in praise of Ontario's manufacturing capacity, but it is to an island in Georgian Bay (where acid-stressed fish populations are declining) that he too retreats.

The full value of this near-northern wilderness, for all its commercialization, over-development and exorbitant pricing, exceeds dollar statistics. Consider Gord Mewhiney, spokesman for the Federation of Ontario Cottagers Associations, in his summation to the Legislature committee in 1979: "What is happening in front of our cottages? We were once told of water pollution, then we were told not to eat the fish because of mercury, and now we are told that our lakes simply don't have a hope in hell. Our northern area, our lifestyle, it's all jeopardized! Put a plug on acid rain, now, before it is too late." But Mewhiney also spoke of the lakes in terms which could not fail to move even the most detached urban politician. "We are 300,000 cottage owners, a million voters, and our property values and future for retirement are at stake here. We have a large economic impact—if we are forced to give up and go south for our relaxation, the cottage area business won't exist within three decades."

The Treasurer of Ontario may offer multi-million-dollar enticements to automobile manufacturers to build factories in the south, but he is also the first to face delegations of worried resort owners from Muskoka. After all, he got his start operating a Muskoka resort, and still does. Muskoka is his constituency. He knows the value of the industry. And his counterpart ministers at the provincial industry and tourism department know it; they've loaned more than $56 million to the industry since 1966 to build it up. By 1979 tourism loans, particularly for resort rejuvenation, were taking up 20 per cent of all annual business development loans. The industry and tourism ministry boasts that its good works are "synonymous with economic growth" in Ontario, and projects that by the year 2000 "tourism will be Canada's leading contributor in income, employment and export earnings." Under existing policies acid rain will continue to fall unabated until the year 2001.

Ontario's environment minister Harry Parrott once tossed off an unexplained estimate of $500 million per year as a possible acid rain damage cost, when he spoke before the Legislature committee. But the economic effects of acid rain on tourism could, almost like the physical effects of acid rain, begin unnoticed and be attributed to something else. Ontario tourism operators objected to newspaper reports about acid rain in the late spring of 1978 as the worst possible publicity for the beginning of the tourist season. Who can blame them? And yet, future seasons may never happen if the conspiracy of silence continues. Dr. David Schindler of the federal Freshwater Institute feared for the worst when he told the committee in early 1979: "There's been much talk of the jobs lost if major polluters are forced to curb their emissions. But I hope somebody is thinking of the thousands of tourist operators who will be out of business in 10 to 15 years if there are no controls on emissions." To 1980, no one in the Ontario government has given that possibility much thought, at least not in any recognizable form of studies, projections, or public warnings. And yet the same government concedes that without an early end to acid rain, the death of the lakes which support that industry is more than a possibility. It is virtually inevitable.

Ontario's aquatic resource base for fishing and tourism is

not the only one at stake. Quebec, the Adirondacks, Minnesota and Michigan all draw substantial revenue from the lakes and rivers. And in Nova Scotia, where it rains acid too, much of the famed Atlantic salmon fishery has already gone down the drain. Federal fisheries biologist Walton Watt explained the situation in an interview in late 1979. With acid levels averaging below pH 4.8 in 1978, the waters of seven major rivers running southeast across Nova Scotia to the ocean had become death traps for the salmon. Twenty years ago rivers like the Mersey, Roseway and Sissiboo provided record catches of the big fish. "Now they're dead. There's no salmon. They're wiped out." He estimated 6,000-7,000 salmon had disappeared, and another 20,000 in other rivers were heading for extinction.

As Watt pointed out, a sports fisherman spends roughly $150 on equipment, travel and lodging when fishing the southeastern rivers. The area had been famous and drew fishermen from far beyond the local region. But now, with more than 6,000 fewer fish, Watt calculated that a minimum of $600,000 was being lost as income in the area, year after year. He admitted his calculations were rough, but for him they had particular meaning. "I know the sports fishing industry has drastically declined. My family used to be part of it." There's no profit now for south-eastern Nova Scotia in sports fishing rivers which today hold nothing but plentiful supplies of eels.

Further south, in the state of Maine, there's mounting concern that a $60 million effort to restore salmon to rivers may be acid-wasted. Beginning in 1966, federal and state authorities began removing unnecessary dams and halting industrial pollution on the Connecticut and White Rivers, in hopes of rebuilding salmon stocks which disappeared nearly 100 years ago and are now considered extinct there, as everywhere in the continental US. In 1978 the first few fish were implanted in the river—almost all died before they could spawn. In 1979 only 60 adult fish made the return migration to the headwaters; 36 stayed alive long enough to lay eggs. But as fisheries biologists warned in early 1980, with acid levels in the White River at the pH 4.8 level, there is little chance the eggs will be successfully fertilized, or hatch. In effect, the rivers had been repaired and cleaned, only to be poisoned again. "We're concerned. Our waters seem susceptible to acid rain," one biologist said, apparently surprised. US Fish and Wildlife Service researchers have begun testing waters throughout the river-shed for acidity, in hopes of finding "how far the acidity has to go before it endangers fish." "If acid levels continue to increase, it will jeopardize the salmon restoration program," lamented Maine researcher Andrew Stout, in March, 1980 at a meeting of the National Wildlife Federation at Butler University, describing the loss as "very substantial."

Nearly 5,000 miles to the northeast across the Atlantic in Sweden, there are details on exactly what is at stake in waters washed with acid. It's the kind of information which has made Sweden a world leader in acid rain research and made the economic realities of acid rain understandable to every Swede. And that in turn has fostered political decisions which have already produced reductions in Sweden's pollution emissions. In a 1978 study,[4] for example, it was calculated that Sweden faces a $16.5 million annual loss in inland commercial fisheries due to acidified waters. Sports fishing and tourism losses total another $50 million annually. Coastal fishing for species which spawn in freshwater (like Atlantic salmon) wasn't calculated in that report, beyond a note that 85 per cent of the migration to the sea would be affected. The report was based on the assumption that the productivity of fish "sooner or later in principle will be zero if the deposition of acid continues at the present rate." And, as the report dispassionately concluded, in the most heavily acidified regions "the disappearance of incomes from fishing and tourism can jeopardize the possibilities for the people to exist and make a living in these areas."

In 1971 the Swedes also considered the damages to forests, property and human health, in dollar terms. As the authors of the "red book" summary readily admitted, absolute proof was not then available, but based on a decade's research the Swedish government felt justified in some shocking conclusions. In forest growth, "an annual rate of reduction amounting to .3 per cent would probably be of the right order of magnitude," the red book reported. And based on the unfortunate likelihood that acid rain source emissions, particularly from Europe, would remain the same as had been measured in 1965—a condition the Swedes admitted "is thought to be the most realistic"—by the year 2000 Sweden will suffer a 13 per cent decline in forest growth. "A direct estimate in monetary terms (of such a condition) hardly does justice to the nature of the damage," the report added, but "the most informative figures are probably that 7 per cent of the raw material base of the country's forest and pulp industry will have disappeared by the year 2000." And that would equal a minimum cost (loss) of $40 million per year, by the year 2000.

The Hubbard Brook Forest, N.H. investigation revealed a 20 per cent drop in wood production in the trees after 1961. The October 1979 first Canada-US assessment of acid rain warned that forest degradation could have reached an irreversible state long before a 20 per cent loss in wood productivity was absolutely proven. And there are other studies. Consider what is at stake, in Canada alone. Trees cover 35 per cent of the land. Directly or indirectly they provide one out of every 10 jobs in the country, $18.5 billion worth of shipped material and $9 billion in added value in 1978, a $10.6 billion in net contribution to Canada's balance of payments in 1979. Canada is the world's leader in newsprint production and export (half the world's total); second in pulp production (Sweden is fourth), and harvests nearly 5 billion cubic feet of wood per year. That's only part of the story. After a century of uncontrolled decimation, the Canadian forest industry is now running out of trees. Massive clear-cutting, token replanting and negligible interest in modern harvesting and silviculture has driven the loggers, mills and plants into increasingly remote areas in search of large forests. What's been cut has too often been abandoned to an agonizingly slow 60-80 year regrowth period. Nearly 12 per cent of the prime forest land is not adequately stocked with harvestable trees now, and 500,000 acres more are added to the backlog each year.

The worst of this forest short-fall exists in eastern Canada, in Ontario, Quebec and the Atlantic region, which provides almost exactly half of Canada's forest productivity. The industry, and more recently the federal and provincial governments, claim they can increase the harvest from this already depleted half of the country by another 40 per cent within 25 years by using improved techniques.

But what nobody seems to have considered at any length, judging by the public statements, is the risk that the forests won't grow, because of acid rain. As the joint Canada-US acid rain assessment in 1979 pointed out, "much of the most productive forest in northeastern Canada lies within the zone most affected by long range transport and acid deposition." The trend to taking more trees leaves the land barren and exposed to acid rain and snowfall, and to slow, irreversible poisoning of the soil. As the Canada-US report noted, acid pollution on merely a fraction of the scale of that typical around the well-known sulphur sources such as Inco's Sudbury smelter *could* produce "the most disastrous consequences for the well-being of a vital resource industry."

If the Swedes are right, or even reasonably predictive, such conditions *will* produce most disastrous consequences for the well-being of a vital resource industry, and indeed the whole Canadian economy.

An almost entirely unnoticed study by the National Research Council of Canada in August 1977 estimated the direct annual loss to forests due to acid rain at that time "to lie between $1.2 and $2.8 million." That study was heavily criticized by provincial governments as unsubstantiated and likely erroneous. And yet, to 1980 further Canadian research into the question has been negligible. The Canadian Forestry Service, which federal cabinet minister John Roberts promised in March 1980 would play "a prominent role in determining the effect of acid rain on forests," is now operating on a budget 50 per cent smaller than it was in 1973. And Roberts, in a major policy speech to the Canadian forests industry in March, 1980,[5] did not hold out much hope for any substantially improved funding. He did however promise the forest industry that the Service would help find better ways to harvest trees. What the Forest Service has yet to find out is whether there will be enough trees to cut, if acid rain continues unabated. (Roberts, incidentally, is also the new Canadian minister of the environment.) There is indeed much at stake, including the federal minister's credibility in professing to worry about acid rain while ignoring its implications for forestry, and Canada's economy.

More than forests are at stake. Agricultural crops are worth $8.9 billion per year in Canada. There is strong evidence that what grows in the soil can be and is affected by the acid rainfall. The extensive experiments by the US Environmental Protection Agency at Corvallis, and others at Oak Ridge, Hawaii and the experimental tobacco plots north of Toronto all show damage to crops such as radishes, beans and tobacco. By the end of 1980 the Corvallis experiments may indicate exactly how little acid rain is necessary to blight, corrode or kill a crop. Matching direct damage with sub-soil degradation may take years of research before precise answers are available. For now, the only sure conclusion is that untold millions of dollars worth of agricultural crops are at risk. And yet, to date there are no Canadian or American policies to avert this risk. It wasn't until October, 1979, that either country formally acknowledged that "there is every indication that acid rainfall is deleterious to crops."

Not all of the economic calculations of acid rain damage need be based on too-few studies and "reasonable predictions." There is one aspect of the damage which is clear cut now, and enormously expensive—property damage. Long before Sweden concluded that sulphur dioxide and associated air pollutants were costing $20 million per year in metal, stone and wood corrosion, engineers and scientists around the world had charted rates of air pollution damage.[6] To cite only one example, Cleopatra's Needle, the stone obelisk moved from

Egypt to London, had suffered more deterioration in the damp, dirty and acid atmosphere of London in 80 years than it had in the preceding 3,000 years of its history. Cement, concrete, metals, paints, even fabrics are victims—flags fade faster and are tattered sooner in cities such as Los Angeles or Chicago than in cities of cleaner air. In 1978 the US President's Council on Environmental Quality estimated that property damage due to acid rain is $2 billion per year. In 1977 the National Research Council of Canada reported that sulphur emissions in air cause an estimated $285 million damage per year in building deterioration, including $70 million in exterior paint damage alone. The distinction between direct damage by air laden with local sulphur dioxide and wind-blown sulphuric acid from distant sources is not clear yet, but the total damage due to sulphur emissions, in one form or another, is obvious. As the International Joint Commission reported in 1979, 50 per cent of the corrosion of cars may be due to acid rain.

On the west coast in traditionally-polluted Los Angeles, Environmental Protection Agency researchers came up with a dollar value for pollution-related visibility, in 1979. As part of a $600,000 study, they found that potential house purchasers were prepared to pay more for homes which had a clearer view. Improved straight-ahead vision across Los Angeles was worth $350 more per house, according to citizens interviewed. The same study also found that in six pairs of neighborhoods identical except for air quality, those in clean air neighborhoods had substantially higher market value. The researchers concluded that a 30 per cent improvement in air quality would increase real estate values by $500 per household, or $950 million per year in Los Angeles. It was a novel attempt to put a price on air, and L.A. is notorious for pollutants other than sulphur dioxides. But how soon will sulphuric and nitric acid haze have reached equally visible expensive levels on the east coast?

And finally, there are health costs. Hamilton estimated at least 5,000 Canadians may die each year because of acid rain-related sulphates; Mendelsohn and Orcutt put the figure at 187,000 deaths per year in the US. The dollar value of a life in

North America is inestimable on an individual basis; on a national basis economists and health care professionals estimate that premature death causes an average $80,000 loss in income alone.[7] As the professionals are the first to admit, such calculations are "highly insensitive" to such factors as housewives who don't earn income, and medical costs, and lost income not directly attributed to illnesses leading to death. But the US EPA, by calculating time and productivity lost, and hospital and compensation costs, estimates that air pollution is costing the country more than $10 billion per year. The US Presidential Council on Environmental Quality estimates sulphur dioxide alone causes $1.7 billion worth of health care costs each year in the US. The costs will almost inevitably be higher by the time research has pinned down exact totals.

To date almost the only discussion of acid rain in economic terms has focused on the cost of turning it off. In surprisingly short order governments and industry have been able to project frightening multi-billion dollar expenditures needed if acid rain sources are to be curbed in North America. One recent (1979) prediction runs close to $5 billion over a decade for Canada alone. There's much to suggest that such sums are exaggerated, based as they are on traditional technology and corporate analyses, but much more importantly, these costs are grossly deceptive. They loom as large as they do—and cast shadows of galloping inflation, corporate bankruptcy, job loss and skyrocketing consumer prices—because they stand alone. The other cost of acid rain—the economic damage—has been left unconsidered. The task of calculating acid rain damage is admittedly far more intricate than the abatement estimates, and governments and decision-makers have so far chosen to ignore the question because they don't know its exact dimensions. There is already some evidence, beyond the work of individuals like Jerry Liddle, which indicates that the cost of acid rain is so enormous it could undermine the financial stability of entire regional and national economies. And while the detailed damage reports are still unavailable, there is much to be gained by recognizing the enormity of what is at stake. No-one knows exactly what acid rain is already costing North America or will cost if the problem increases as presently planned. Already, costs are enormous. Future costs, however, go well beyond the merely economic. As biologist Tom Hutchison of the University of Toronto told the Ontario Legislature committee,

> deterioration of our lake environments, of the fisheries and the recreational aspects that go with it, is going to hit a lot of people very hard. ... The average man in the street, certainly in Canada, has an enormous respect for the environment. If we allow our short-term solutions to problems to devastate that environment, as we are on the way to doing now with acid rain, I think we are going to have to do a lot of answering to a lot of people in 15 to 20 years time.

Unfortunately for Hutchison and that environment, 15 years time is beyond the normal vision, and term of office, of those making the decisions now.

# 6
# Weak Laws and Big Bucks

When the Ontario Legislature committee investigating acid rain ended its inquiry in February, 1979, it had accumulated nearly 1,000 pages of transcripts and supporting documents. The inquiry represented the most detailed public assessment of the acid rain question to that date, and did much to awaken Canada to the dangers. Yet the committee came to a surprisingly brief conclusion. There were only two options for acid rain: turn it off, or try to live with and repair the damage. The committee correctly decided that repair was "neither economically nor physically feasible." Repeatedly liming acidified lakes—the only repair-cost figure the committee was given—was estimated at $150 per acre, too expensive in a province with hundreds of thousands of acres of acid-vulnerable water. Nothing was said about costs to the tourist industry, forest productivity or a chemically over-dosed agriculture. Even so, it was clear that the only hope was prevention and reduction of pollution.

Recalcitrant Conservative government members of the committee acknowledged the need to reduce emissions and the Opposition members insisted on it, vociferously. It came as a surprise to the members to be told, by federal—not provincial—environment ministry officials, that "the technology to ensure a huge reduction of sulphur dioxide is available now." It came as no surprise to be told that applying such technology would be expensive—Inco Ltd. officials suggested its costs would range beyond $500 million towards $1 billion, and implied such costs were simply unacceptable. And as Inco put it, what's not financially feasible is not technically possible. What also came clear in testimony before the committee, and was in fact already self-evident, is that there are no specific laws in Ontario which could immediately impose abatement technologies on the principal sources at any cost. The same applies in the United States.

The problem with acid rain is that it extends beyond the laws; it is a regional, national and international pollution while the laws are almost exclusively local.[1] What goes up the stacks and comes down 500 miles away is ignored by local regulations; what comes down from the sky on a region 500 miles away is almost an Act of God. To make matters worse, in both countries there are no laws which relate to acid rain itself. The only limit is on sulphur dioxide—and a little sulphur dioxide from a lot of local sources equals acid rain over a huge region. The latest sulphur dioxide limitation was imposed in Canada in 1976; in the US as recently as 1978, before acid rain was acknowledged by the technocrats. It is fully recognized now, but remains unregulated.

In Canada the federal government is supposed to call the play, from the sidelines. Ottawa proposes limits on sulphur dioxide; it is up to the provincial governments to apply them. The feds have declared that sulphur dioxide should not exceed .02 parts per million in air, averaged over a year, for the sake of protecting public health. But although Ottawa has the constitutional authority to impose this standard, it conspicuously lacks any mandatory responsibility to enforce it, and does not. Ottawa's is not a law on how much sulphur dioxide can be emitted by any source, but only a law saying the regular or "ambient" air should not be polluted beyond that .02 level. How the sources are controlled is up to the provinces. They make the laws that matter.

Ontario adheres faithfully to the federal maximum of .02 parts per million of sulphur dioxide in air, and enforces it, whenever the environment ministry chooses to. This is supposed to ensure that no one suffers direct health damage from sulphur dioxide. But for acid rain the concept of ambient air regulations is an awkward one. It's akin to insisting a local garbage dump should have no more than two tons of sulphur in it, without saying how much sulphuric acid anyone can dump there. When it comes to regulating the dumpers, Ontario negotiates separately with each one. There are no uniform laws on source emissions. As long as a source cannot be proven the direct cause of violations of the ambient maximums, it is free to pollute. In a specific sense, the only regulation in Ontario's Environmental Protection Act concerning sulphur dioxide only requires that at the point the emissions are measured—not at the source—they must be no more than 14 times the normal "safe" pollution level in a 30 minute period. This regulation includes 29 pages of details on calculating the link between a source and "the point of impingement"—a cumbersome law which would deter most people from bothering to measure the dumping of sulphur dioxide. Also, it doesn't work for acid rain.

It's long been claimed in the pollution-control world that "the solution to pollution is dilution." That solution and Ontario's regulations solved Inco's local responsibilities, by adding 3,600 tons per day to an atmospheric dump for sulphur dioxide which now measures hundreds of thousands of square miles. Technically, the air is not overloaded with sulphur dioxide, but it literally overflows with acid rain. The dilution solution— building tall stacks to blow sulphur dioxide elsewhere—has been the cornerstone of Ontario's air pollution legislation for years. It remains so.

The federal and provincial government both make no distinction between emissions from existing sources and new ones. A new plant can be as dirty as an old one. Atikokan, an $800 million coal-fired Ontario Hydro power plant, is under construction now without any better pollution controls than the five other existing coal-fired plants in the province. They emit 450,000 tons of sulphur dioxide per year, through tall stacks. Because Ontario law makes no new-source control distinctions (based on the old assumption that the air is an acceptable dump), Hydro was allowed to proceed without the extra $75 million for proven technologies to cut Atikokan's emissions by at least 50 percent. Atikokan did, however, plan a 650-foot smokestack, to carry those 100 percent emissions far away to join the continental acid rain mass.

In the United States, air pollution laws are more specific, in some cases, but the ultimate result is the same. Unlike Canada, the US federal government sets the important laws, and requires the states to enforce them. A national health-protection ambient maximum of .03 parts per million (similar to Canada's) exists everywhere. Unlike Canada, the US also has an environmental protection standard, of no more than .5 parts per million, in a three-hour average, although the standard is not enforced. But ambient standards are easily met by pushing local emissions up tall stacks and far beyond the local area, the region or even the state. And the national sulphur dioxide standards do not include the sulphates and acids which come down somewhere else as acid rain, although the US Environmental Protection Agency began considering the need for sulphate standards in 1980.

In 1971, amid national alarm over visibly worsening pollution, amendments were made to the US Clean Air Act. The federal government began pressing individual states to crack down on sources, particularly power plants. States which refused to file their plans for abatement initiatives were penalized by reductions in federal funding (including, ironically, water pollution control grants). How the states achieved this reduction was left up to them. They left it up to the emitters. Burning low-sulphur coal was an option, or coal "washing" (treating it to remove the sulphur before burning), or installing devices to "scrub" out the sulphur from the smoke before it left the stack. Many states simply approved the polluting industries' first and easiest solution: the extension of tall stacks to float the problem past local pollution monitors.

Out of sight, out of mind—but it worked, on a short-term basis. In a pitch similar to Inco's hard-sell for the 1972 Superstack the American Electric Power Company took out newspaper ads in 1973 bragging that it was a "pioneer" in the use of tall stacks "to disperse gaseous emissions widely in the atmosphere so that ground level concentrations would not be harmful to human health or property." The company claimed that gasses from the tall stacks "are dissipated high in the atmosphere, dispersed over a wide area, and come down finally in harmless traces." The company went on to blast "irresponsible environmentalists" who argued for tough emission standards at the source of the pollution, charging that they were guilty of "taking food from the mouths of the people to give them a better view of the mountain."[2] By 1976 there were more than 200 smokestacks exceeding 400 feet in height. In 1977, the US EPA had to impose limits on stack heights, but the damage had already been done. In January 1979, the EPA reported that nearly 100 million Americans were living in areas where ambient standards for pollutants (not exclusively sulphur dioxide) were not being met. The majority of those people lived east of the Mississippi, towards the seaboard where air-blown emissions from the coal states accumulate.

Yet in 1971, the EPA had declared a major new offensive on air pollution. Backed by the Congress, it declared that all new polluters must do better than existing ones. In a sophisticated strategy, based on the ever-American belief in inevitable obsolescence and improved replacement, the Agency ordered all new coal-fired power plants to produce smaller emissions of sulphur dioxide, equal to using the best available technology on the market, of no more than 1.2 pounds of sulphur for every million units of heat energy produced. The limit was based on the ability of scrubbers (smoke-stack cleaning systems) which at the time could reduce by 70 per cent the sulphur dioxide emitted from coal and oil containing 3 per cent sulphur.

In 1977, as required by legislation which makes it review the Clean Air Act every 5 years, the EPA began drafting even tougher limits on pollution from intended new plants. An 18-month battle erupted in political and business circles, with the coal-burning utilities protesting that technology was still unable to meet the old 1971 law, let alone any tougher version. To 1977, some emissions had been kept down by simply burning cleaner coal. But less than 20 percent of eastern US coal is "clean" (below 2 percent sulphur) and the economics of using plentiful low-sulphur coal from the western states was declared, by eastern-based industry, to be unacceptable. A second option called coal "washing" involved crushing the coal and chemically separating out the sulphur locked in combination with other rock, but less than 14 percent of reserves were suitable for washing, the coal industry warned. Inevitably the debate turned to the emission control technology the EPA had set as ideal and available: sulphur dioxide scrubbing.

Basically, the principle of scrubbing is as old and simple as high school chemistry. Sulphur-laden fumes and other gasses from burning coal, oil or other materials are blasted through a bath of water and chemicals. The chemicals attract and combine with the sulphur, to fall out as a wet sludge, while the sulphur-free scrubbed gasses pass on up the stack. Half a dozen common chemicals can be used to "scrub" the sulphur, including sodium, magnesium, lime and limestone (the same material used to neutralize acidified lakes). Limestone has a particular advantage—when limestone-sulphur sludge is sluiced off and partially dried, it can be processed into gypsum, a common building material. More often, the unusable sulphur-sodium or magnesium-sludge ends up in giant holding ponds, abandoned mines and quarries.

To handle sulphur-laden, 250-degree hot fumes blasting out of a typical 250 megawatt coal furnace, the equipment of a scrubber plant is massive and intricate beyond any chemistry laboratory—50 tons of water per hour, 120 tons of limestone per day, miles of piping, conveyors, and air ducts, inside a typically five-storey building adjacent to the coal furnaces. But it works. Even before the US government began clamping down on sulphur dioxide air pollution in 1970, scrubber technology had reached the state of cleaning out 70-75 per cent of sulphur dioxide from coal plants. And beginning in 1971 a handful of plants were forced to apply the scrubber technology as their only means of cutting emissions to avoid violating the toughened ambient air standards. By 1978 the techniques had advanced to a 90 per cent or better scrubbing capability.

But throughout the 1970s utilities had castigated scrubbing as an unproven unreliable technology. Only a handful of power plants had adopted it with what the utility industry claimed was limited success. It required huge adjacent plants and risks that a jammed conveyor belt or plugged water pipe could render the entire plant inoperative or shut down power generation. There were even complaints that the left-over sludge presented a serious environmental and aesthetic blight on the landscape. Industry preferred to build tall stacks and push the air pollution out of sight. In a country which had managed to rocket to the moon and harness the power of the atom itself, the claim of technological unreliability was odd, but effective. Ontario Hydro was still using it in late 1978, when it argued that one reason for refusing scrubbers at Atikokan was their "not proven reliability."

When the EPA finally confronted the criticism of scrubbers head on, it turned for help to the masters of applied technology: the Japanese. As a government-industry task force discovered in a February, 1978, tour of Japan, scrubbing technology has been the key to a 50 per cent reduction in sulphur dioxide air pollution there since 1967—despite a 120 per cent jump in energy consumption.[3] By 1978 more than 500 scrubber plants were in operation in Japan, on average removing 90 per cent of fumes from coal and oil-fired power plants, smelters and sulphuric acid plants. Ironically, much of the technology and in some cases the actual equipment had been supplied by US firms. It was being applied to both new and existing plants. At the typical Electric Power Development Company's Takasago plant, scrubbers were added in 1975 using a US-developed technology. To 1978 the pollution control devices were removing 93 per cent of the sulphur dioxide fumes without failure 99 per cent of the time. It had cost the utility $42.6 million (US) to clean up; it cost only 12 per cent of the price of the electricity produced to keep it clean.

"Flue gas desulphurization [scrubbers] is working well in Japan," the US task force reported back to Washington. "But ... a number of differences became apparent during the comparing of the Japanese situation with that of the US." Japan produced a usable gypsum byproduct from almost all of its scrubber plants, the US did not. Scrubber plant employees were specially trained to ensure permanent reliability, US utility employees were not. Japanese industries face a continually escalating sulphur emission tax (exceeding $4,000 per day in 1978) regardless of their control successes, to encourage them to do better. The tax is used for medical care of air pollution victims. Nothing similar exists in North America. And finally, the US investigators reported, "The last US-Japan difference can best be expressed as a sincere co-operative spirit which appears to exist among Japanese industry and the regulatory agencies. ... The national goal of a cleaner environment has been accepted by the utility industry which has made a sincere effort to buy the best scrubbers and operate them in the manner for which they were designed."

Citing the Japanese task force report and refinements in scrubbing technology (which reduced the chemicals and wastes), the EPA in 1979 finally convinced Congress to tighten up the pollution limits on future coal-fired power plants. The limit set—between 70 and 90 percent reduction in acid rain-causing fumes, depending on the dirtiness of the coal—represented less than scrubbing technology could achieve, but at least it put an end to the technological blackmail waged by industry. The tougher new-source emission limits were clearly needed, for the country was desperately turning to coal as a replacement for costly oil. Up to 350 new coal-fired power plants will be needed by the year 2,000, according to the US department of energy. And, if the EPA new-source performance standards are consistently applied, every one of those plants will be emitting 12 pounds of sulphur dioxide per ton of coal, instead of the existing plants' 160 pounds.

The cost of installing scrubber systems as these new plants are built is not, at first glance, cheap—it will add nearly $10 billion to the capital cost of the coal-plant expansion program. But as the US EPA pointed out in considerable detail in May, 1979 when the legislation passed through Congress, the cost is minute compared to the total capital cost of the program, estimated at an incredible $770 billion between 1979 and 1995.[4] Even in America, such sums are almost beyond comprehension to the average citizen, but when it comes to energy, America is going for coal in a big way. Pollution controls are a small price to pay, the EPA declared; they will increase electric utilities' annual revenue requirements by only 2 per cent, which in 1995 will translate into only a $1.20 higher monthly electricity bill for residential consumers. Indirect costs to those consumers, passed on by industrial electricity users, will add another $1.20 per family per month, in 1995. As the EPA's administrator Doug Costle declared, in May, 1979, Americans could certainly afford air pollution controls on their new coal plants.

However, the EPA failed to point out two critical facts. Even with the much-praised controls, the total emissions from all the predicted new plants to be built to 1995 still represent a hefty increase in US emissions. Much less sulphur dioxide from each single new plant, multiplied by 350, equals more overall pollution—an estimated 2 million tons per year. The new standards, with all their compromises, will end up increasing sulphur dioxide rather than holding it constant, let alone reducing it. The new controls would herald a reduction in total emissions if the plants were being developed only as replacements for old ones. But within this century, this will not happen. By 1990, according to both EPA and utility industry projections, existing dirty plants will still account for 80 per cent of the sulphur dioxide emissions. In other words, another 20 years of coal burning and tall stacks under almost no restrictions. The assumption that early obsolescence would phase out the dirty plants, a cornerstone of EPA policy, is already breaking down. Because it is cheaper to operate the old, uncontrolled plants, utilities in the US are already gearing up to use them as constant power sources, with the new ones in reserve for peak demand. Scheduled phasing-out of old plants is being scrapped for repair programs and extended, dirty life. Somewhere after 1995, when binder twine and baling wire can no longer hold them together, the plants will begin to close, and total emissions of US sulphur dioxide may begin to decline. But until then, under existing regulations, the emissions will increase, and with them, the spread of acid rain.

There are no laws requiring pollution scrubbers on the old plants. The utility industry says it will cost too much. Aside from vague references to gradually mandating cleaner coal for old plants the EPA has not even attempted legislation to impose scrubber technology, the only real solution, on the more than 250 old plants in the US. The utility lobby in the US, aided by coal interests, warns that existing plants are financially uncontrollable, citing for example a $1 billion cost predicted for reducing the Tennessee Valley Authority's 14 coal-fired power plants' 2.1 million tons per year total emissions by 40 per cent. After TVA comes American Electric Power Company with 14 plants and 1.5 million tons per year emissions, and Commonwealth Edison with 13 plants and 632,000 tons per year. The emissions from the 10 plants in the US which emitted the most sulphur dioxide comes to 7,900 tons per day, or almost 3 million tons per year.[5] All of those plants lie in the eastern third of the US.

According to preliminary estimates the EPA says it would cost an average of $100 million to apply pollution scrubbers to a typical 500 megawatt coal-fired power plant. On vastly larger single sources like the TVA's Muhlenberg, Kentucky, plant, which emits 1,075 tons per day of dioxide, the cost would be higher.[6] Some of the TVA's electricity customers have already begun court fights to stop plans to reduce emissions from the TVA's coal-fired power plants. They object to the anticipated increased cost of power. These will be interesting cases to watch, since the US government owns the TVA.

In Canada, however, there are no anticipated hundreds of new coal-fired power plants which need strict emission controls.

There are in fact less than a dozen such plants on the drawing boards, across the country. Nor are there 250 existing coal and oil-fired plants still belching virtually uncontrolled emissions in eastern North American skies. Electricity in Canada comes foremost from water power, and provinces such as Quebec, Manitoba and British Columbia hold still untapped reserves. Coal, oil- and gas-fired power plants in Canada number less than 40. Admittedly none of those plants anywhere in the country have tight sulphur dioxide controls, but as a total they emit only 10 per cent of the total sulphur dioxide air pollution in the country.

The only significant utility polluter is Ontario Hydro, which at 450,000 tons per year from three principle power plants ranks as second-worst emitter in Ontario, behind Inco. Not unlike its US counterparts Hydro resists application of scrubber technology, using the same arguments of cost and fallibility. And unconstrained by any new source regulations, it is proceeding with construction of a small, but politically important new power plant at Atikokan. Insignificant by comparison with Inco's vastly larger Sudbury emissions, Ontario Hydro pollution has remained largely ignored, by both Hydro and the Ontario public. The Ontario government owns Ontario Hydro.

In Canada the problem sources are not the power plants, but the smelters, like the copper and nickel smelters of Quebec, Ontario and Manitoba which emit just over 50 per cent of all the sulphur dioxide pollution in the country. Noranda Mines' Murdochville, Quebec, smelter is under no direct orders to reduce its emissions of 96,000 tons per year. The same company's much larger smelter at Rouyn-Noranda, Quebec is equally unrestrained in its emissions of approximately 605,000 tons per year, plus 443 tons of arsenic, 210 tons of cadmium, 11,512 of lead and 2.8 tons of mercury.[7] In an interview cited in the *New York Times Magazine*, on October 21, 1979, Alfred Powis, chairman and president of Noranda, had the following comments on his company's air pollution:

If anyone could ever demonstrate to us that we were doing

serious damage to the environment ... then we would consider shutting our operations down. We have never been able to find any evidence or proof of such damage. We operate with a clear conscience. We can't even find evidence that we have damaged vegetation in the area. In the timber forests they have never found a dead tree—or group of trees—from sulphur dioxide. Our emissions are carefully monitored by ourselves and the Canadian government.

Powis offered his corporate clean bill of health less than a year before the Quebec government released its first of 36 volumes of detailed studies on the emissions of Noranda Mines Ltd. The studies and summary report contradicted virtually every inference and factual claim of the chairman of Noranda. The summary noted that a group of Rouyn-Noranda citizens successfully sued the company in small claims court in 1977 for damage to local vegetation from the smelter's emissions. It further stated "sulphur dioxide is responsible for the deterioration of vegetation in the area. ... [In the north part of one lake] levels of copper and zinc exceed by 100 times acceptable limits." It concluded that "a marked increase in concentrations of sulphates in rainfall near the town was noted, with the maximum being reached several kilometres away in the direction of the emission cone of the [Noranda] chimneys." Noranda's tallest chimney-stack is 500 feet.

The Quebec study, first begun in 1977, was limited to environmental conditions close to the smelter town. "The range of sulphur dioxide and its effect on the acidity of rainfall are felt over much greater distances (up to 1,000 kilometres)," the report pointed out. But "rainfall in this part of Quebec is acidic (pH 3.6-4.0)." It also added that although Noranda's annual sulphur dioxide emissions remained below the maximum safe level for health effects, in the immediate area, wind currents repeatedly carried clouds of this and other pollutants throughout the community at levels far in excess of the safe level. "It is reasonable to associate excess mortality due to respiratory problems to the significant concentrations found at certain times in the ambient air of Rouyn-Noranda." The

Quebec study noted that sufficient health records were lacking, but more research is under way. The government of Quebec has only begun to press for emission abatement at Rouyn-Noranda. And it is fair to suggest that the chairman and his company have only begun to resist, although the ultimate employment threat has already been voiced: "If anyone could ever demonstrate to us that we were doing serious harm to the environment...then we would consider shutting our operations down."

About 150 miles southwest, across the Ontario border, lies Sudbury. To 1980 the Inco Superstack was licensed by the Ontario government to emit 3,600 tons per day at least until 1983. An undetermined reduction was presumably to be ordered, after that date. In 1970 Inco had been ordered to cut emissions from almost 6,000 tons per day to 750, by 1978. The company failed to get below 3,600, and in 1978 was given to 1983 to continue at that rate. Across town in Sudbury lies the other, almost forgotten smelter: Falconbridge Nickel Mines Ltd. Less than one-third the size of Inco (1979 earnings of $130 million, 2,000 employees in Sudbury and capable of producing 80 million pounds of nickel per year from its mines), Falconbridge in 1978 completed an $83 million rebuilding of its smelter which cut its sulphur dioxide emissions 80 per cent from those of a decade earlier, to 460 tons per day. The company converts its sulphur dioxide to sulphuric acid, which it sells. Falconbridge faces no major environmental control orders by the Ontario government.

Another 1,000 miles northwest, at Thompson, Manitoba, Inco (again) operates its newer nickel smelter under provincial emission limits of 1,250 tons of sulphur dioxide per day. (The smelter was running at much less than capacity in 1980 and putting out an average of 650 tons per day.) It faces no orders to reduce emissions. A few miles east of the Manitoba-Saskatchewan border, Hudson's Bay Mining to mid 1979 was under no orders to directly limit its emissions, but only urged to stay within the provincial and national ambient limit of .02 parts per million, on an average annual basis. The smelter was thus operating at about 800 tons of sulphur dioxide per day, and expected to receive a renewed license to continue at this level, in 1980.

These then are the major, specific sources of sulphur dioxide in Canada. With good reason has Inco's Sudbury smelter been the focus of most attention and debate over the necessity and cost of abating its emissions. It is, after all, the world's largest smelter, and North America's if not the world's largest single source of sulphur dioxide. The company's record of abatement and the response of government to the super-problem of the Superstack are in many ways representative of the typical machinations between government and powerful polluters, and Inco's case is dealt with in detail elsewhere in this book. In early 1979 Inco had warned the Ontario Legislature committee investigating acid rain that any substantial reduction of emissions at Sudbury "would have to be measured in billions of dollars." As the committee report noted, the provincial environment ministry had pathetically little knowledge of the validity of this claim. But the billion dollar figure, or threat, went on record, largely unchallenged, and remained a powerful disincentive to crack down on Inco, let alone any other major Canadian source. The cost of a national clean-up of sulphur dioxide emissions remained largely unconsidered, yet was assumed to be enormous.

Federal environment minister John Fraser had peppered his speeches with references to this enormous cost of abatement, in 1979. In September he had declared that a 50 per cent cut-back in eastern Canadian sources would cost "as much as $350 million annually [and] a similar reduction in the northeastern US may take $5 billion per year." By November, 1979, at the citizens' international conference on acid rain Fraser had refined the figure, upwards, to $500 million per year for 20 years in Canada, or $10 billion in total, and at least 8 times more for the northeastern US. The $10 billion figure became a major factor in Ontario's argument that neither the province nor the country could rush hastily into a major abatement program with such potential for economic disruption, an old but familiar song in Ontario where pollution controls are at issue.

Big numbers like $10 billion make good newspaper copy, even if they represent only one half of a serious cost-benefit analysis. And lacking any federal or provincial mention, let alone study, of the financial benefits of reduced emissions, that figure has become notorious in the acid-rain debate in Canada. But in fact, that $10 billion figure is strangely vague. As a senior federal environment official conceded later, "It was an initial estimate, only intended to give a feel of the kind of size of estimated cost we thought we were facing. It wasn't based on any detailed study of the technology; no, there is no report or internal government study. It was just, well, I'm not sure where we got it. It was just rough."[8] In a later interview he explained that the figure was an estimate of the cost of abating by 50 per cent all the major smelters in Canada, and Ontario Hydro. He did accept the suggestion that $10 billion for clean-up did seem "pretty frightening, at least in some circles."

Yet in mid-April, 1980, the new federal environment minister John Roberts unexpectedly announced that Inco alone could achieve a 50 per cent emission reduction, by 1985, at a cost of only $400 million. The cost, based on rebuilding the smelter core to produce marketable sulphuric acid, could be borne by Inco without any important financial unease, and without government aid, the minister declared. Roberts' declaration was based on "very reliable" analyses of Inco's financial status and rough itemization of the costs of existing abatement technologies, the first time any government in Canada had taken the time and initiative to consider them. Not surprisingly, Inco chairman Charles Baird dismissed the validity of the federal studies, although he conceded that technology did exist to produce a 50 per cent or better reduction. Baird refused to explain his version of the cost, preferring allusions to "hundreds of millions of dollars" and insisting he did not mean just a couple of hundred million. Even that cost was less than what his company had argued before the Legislature committee only 18 months earlier. Baird did not comment on his firm's ability to pare down its costs, even if only from "billions" to "hundreds of millions." There is every reason to suggest that the cost of cleaning up a handful of smaller smelters across the country and a very few power plants will be equally much less than the original hasty and frightening estimates of $10 billion. Where there's a will, there's a way, as the adage goes. What's lacking in Canada is the political will to point the way for polluters.

One country, at least, has faced its acid rain problem squarely and has applied active, resourceful legislation to curb the damages. Granted, Sweden was one of the first nations to spot the pollutant, much of which swept in on the country from the Continent and Great Britain. As far back as the 1930s Swedish fisheries inspectors reported that extremely sensitive roach were disappearing from rivers and lakes on the west coast. In neighboring Norway in 1926 a researcher named Sunde had already demonstrated that high mortalities in salmon hatcheries were caused by the acidity of incoming stream waters. By 1955 biologists Barrett and Brodin had published measurements of acid rain across Scandinavia and linked them to lake and fish loss. They included warnings that alkalinity was the key to measuring the buffering capacity of the lakes, and that the alkalinity was declining. In 1968 Svante Oden, a biologist at the Swedish Agricultural College at Uppsala, described how acidic air pollution was affecting soil and surface waters, retarding plant growth, changing ecosystems and the biota in the lakes and rivers, killing some organisms, and much more.

Further studies projected a 7 percent loss in forest resources, a staple of the Swedish economy, and concluded that corrosion and property repair from sulphur compounds in the air would rise from a 1970 level of $500 million to almost $2 billion by the year 2000, if sulphur dioxide concentrations remained unchanged. The figure covered only damage to steel and painted woodwork; the costs of erosion of limestone, sandstone, concrete and cement structures weren't tallied.

Sweden's 1971 reports also touched on health. By their own admission, the authors' conclusions were speculative, and Sweden already had mandatory maximum limits on sulphur dioxide in air pollution. But if pollution-linked deaths did occur at lower-than-maximum-permissible levels—"something not contradicted by available data"—then even slightly increased sulphur dioxide and compounds could hold significant risk of increased health hazards, the report warned.

The Swedish government's 1971 report on acid rain was followed soon by a full English translation as Sweden's chief contribution to the United Nations Conference on the Environment, in Stockholm in 1972. In 95 pages summarizing more than 46 other reports covering two decades of research, the Swedes outlined everything known and suspected, about acid rain: sources, volume, impact areas and damage, the cost of the damage and the cost of clean-up and prevention. They included references to thousands of dying lakes in Scandinavia, termed the present situation "disastrous," gave 10 and 28 year predictions of what would happen if the rain continued, and urged a policy of air-pollution control. They also noted that almost identical conditions of geography, weather, vulnerable lakes and acid rain sources probably existed "in Canada and the northeastern United States. A detailed study of the likelihood of such a development [acid death] is a matter of urgency." The Swedish report was clear-cut, comprehensive, and comprehensible by anyone. Published as a small red-covered book, it was ammunition for Scandinavians to launch a war against their acid rain, and within a year Norway and Sweden had drafted policies to turn it off, Norway committing $10 million for research alone. Yet on this side of the Atlantic, even today only a handful of scientists and fewer politicians have ever read Sweden's little red book.

As early as 1969 Sweden had begun to legislate against acid rain by imposing limits on the burning of heavy fuel oil containing more than 2.5 percent sulphur. But by 1975, with more evidence to support the warnings of the 1971 red book, Sweden felt compelled to consider much larger reductions. As the red book had predicted, the country was then putting out almost 1 million tons of sulphur dioxide fumes—80 per cent of it from oil-fired heat, power and electricity plants serving both industry and consumers. The studies had also considered ways of reducing that pollution output, by oil with less sulphur, or devices to filter the fumes and capture sulphur, or laws requiring industry to abandon "dirty" processes. By 1977 the government had weighed the options and acted: the big sources—fuel oil—would be curbed first; they produced the worst emissions. The 16 most heavily-developed Swedish counties, burning 75 percent of the heavy oil, were ordered to buy cleaner oil, thus cutting the sulphur content from 2.5 per cent to 1 per cent. By 1984 the ban will apply to the whole country. Light fuel oil such as diesel oil containing more than .3 per cent sulphur was banned, everywhere in the country, in 1977.

Industrial processes which emitted sulphur dioxide for other reasons were not left uncontrolled, despite their relatively small contribution. By 1985 those industries must cut their emissions by 50 per cent. The technologies or alternative processes already existed to accomplish this, the Swedish reports pointed out, but some time for gradual adjustments was warranted. To ensure the sulphur abatement targets are met on time, Sweden also introduced "economic incentives": sulphur dioxide polluters who fail to meet the deadlines will pay increasingly steep sulphur emission fees calculated to eliminate any financial advantage the factories might gain by carrying on with cheap, dirty processes.

Oil is to Sweden as coal is to North America—a vital source of energy, and acid rain. Sweden consumed nearly 19 million tons of oil for industry, power and heat in 1974, and the aggressive determination to cut back on the cheapest, dirtiest forms of oil (and thus cut sulphur emissions by almost 50 per cent) by 1985 was an expensive one. The policy would cost industry and consumers an extra $115 million by 1985, according to estimates made before the abatement actions were mandated. And yet, these regulations passed easily through the Swedish Parliament. As was clear as early as 1971, the cost of not taking such steps would be far greater by the year 2,000, in damaged forests, corroded property, destroyed fisheries and abandoned tourism.

This decision to reduce acid-rain potential was not, however, without its controversial implications. If emissions are to be reduced in the face of rising demands for energy, a clean alternative fuel is necessary. Sweden opted for more nuclear power. The country's nuclear reactor program was geared to a six-fold increase—at least 12 reactors by 1985. As the national commission which drafted Sweden's acid rain reduction policies in 1976 pointed out, the reactors would provide 66 billion kilowatt hours of electrical energy without the 400,000 tons of sulphur dioxide per year that equivalent oil-fired power plants would emit. The go-nuclear policy was not taken exclusively on behalf of acid rain control—skyrocketing oil prices had at least as much to do with the decision. And the cost of the multi-billion dollar nuclear program was not included in calculating the cost of acid rain abatement. The controversy over nuclear expansion has raged unabated ever since it was taken, in Sweden, toppled at least one government, and to the end of 1979 was far from resolved. The debate now centres around how much energy consumption can be reduced, in total, through conservation and renewable alternatives. No-one, however, suggests the answer is to return to sources of power which produce acid rain.

What makes Sweden's initiatives most remarkable is that they would not cure Sweden's environmental problem, but at best only keep things from worsening quickly. Without European controls on sulphur sources, Sweden will still suffer at least 310,000 tons of foreign, wind-blown sulphur overhead by 1985—almost the amount Sweden committed to cutting from local sources. "Even if stringent controls are imposed in Sweden the ... sulphur will increase if control programs are not instituted in other European countries," a government commission warned in 1976. No such controls in Europe were promised, or even predicted, at that time, and yet Sweden went ahead. Since 1978 Sweden has been the leader in pressuring

Europe for a reduction in its sulphur emissions. Swedes laden with reports, projections and cost-benefit analyses show up at every occasion the issue is on an agenda. In 1979 Swedish scientists disrupted a NATO scientific meeting in Toronto on the subject, their charts fingering Britain for steady streams of sulphur-laden air crossing the North Sea to Scandinavia. Around the headquarters of the European Economic Commission, Swedes are known as both the experts and the agitators on acid rain-control proposals. Sometime in 1980 the EEC is to begin detailed study of a Swedish proposal for an international agreement to make Europe curb acid rain. With clean hands, and comparatively clean air, the Swedes press for action, while the acid rain increases.

Sweden initiated its acid rain cleanup by relatively simple laws which imposed a simple technical fix: reduction of sulphur in the oil used at the power plants. These sources caused more than 80 per cent of the problem (excluding automobiles). But Sweden has few coal-fired power plants, and even fewer dirty smelters, unlike Europe and North America. (In an ambitious estimate in 1976 Swedish researchers predicted it would cost western Europe alone $450 million per year by 1985 to achieve a reduction from 22 million tons to 8 million tons of sulphur dioxide per year.) Oil's ever-escalating cost is already pricing it out of the market for boilers and generators anyway. Sweden is only slightly larger than Newfoundland, with a population equal to Ontario's, and yet its prompt, determined response to the acid-rain problem should shame larger countries. We know what the trouble is, we have the technology to solve it. Curbing acid rain will actually be cheaper and more productive in the long run than letting it fall unabated. What is holding us back?

# 7
# Acid Politics in Ontario

North of Lake Superior and 650 miles beyond central Ontario lies the town of Atikokan, population 6,000, a soon-forgotten stop on the lonely highway linking Ontario and Minnesota. By the mid 1970s the town appeared doomed to be forgotten forever, as the nearby Steep Rock Iron Mine, the main industry, began to run out of ore. As the mine laid off employees, Atikoken started the agonizing slide towards ghost-town status, the usual fate of northern communities subject to resource depletion. A visiting journalist described it this way: "The bank looks shabby, with tape covering a hole in the window made by a vandal. An unfinished foundation is a reminder of optimistic plans. The merchants sit in near-empty shops on Saturday afternoons. Brick-tossing and tire-slashing are the pastimes of youths with nothing to do."

Atikokan's charm may be elusive, but the surrounding region is something else again. About 10 miles south begins the boundary of Quetico Provincial Park, a 1,120,000-acre wilderness of beauty unsurpassed anywhere in Ontario, specifically limited to canoeists and hikers. There are no roads into the interior of the park. As far back as 1909, Quetico had been set aside as primitive wilderness to be preserved intact for all time. That same year the United States government had made similar declarations concerning Superior National Forest, which lies across the border. Between the two forest preserves lies the Boundary Waters Canoe Area, a million-acre maze of rivers, streams and lakes, one of the finest canoe-tripping areas in

North America. Superior is the largest virgin forest in the US, and the most popular recreation wilderness in the country: fifty million people live within two days travel of the Boundary Waters.

In 1976, the government-owned Ontario Hydro proposed to build a coal-fired electricity generating plant near the town as a source of power for the northwest, and equally important, as an injection of employment and revenue into the area. The project would require at least 1,000 construction workers. The plant was to be moderately sized, consisting of four 200 megawatt generators to be constructed at a cost exceeding $800 million and slated for partial operation in 1983. Ontario Hydro's preliminary environmental impact documents showed that the plant and its emissions of 200 tons of sulphur dioxide per day from a 650-foot smokestack would cause no harm to the regional environment, and the project was given a go-ahead by the Ontario government in June, 1977. No detailed, open-to-the-public assessment of the need for the plant and its risks was held.

The Atikokan project did not go unnoticed. Even before its approval, environmentalists from the US had begun to protest the plan, arguing that the uncontrolled sulphur dioxide emissions threatened to blight the Boundary Waters wilderness. By August, 1977, Atikokan had become a political irritant, with the Minnesota government raising formal objections to the Hydro plan. The border wilderness was of such exceptional beauty and fragility that it deserved full protection from any environmental intrusion, said the Americans. Hydro should spend another $75 million to install air pollution scrubbers to remove at least 50 per cent of the sulphur going up the stack, or at least delay the project until all the environmental impact had been clarified. Largely swayed by Hydro's private analyses, Ontario rejected Minnesota's claims, citing "the lack of any indication of any potential damage." According to the Hydro argument, the uncontrolled Atikokan plant was simply too small to add any substantial sulphur dioxide fumes to regional air currents already laden with acid, much of which rose from sources far to the south in the big coal-using states. And the 650-foot stack at the Atikokan plant ensured that no low-level sulphur dioxide would roll across the nearby wilderness to leave tell-tale signs of acidic lakes, blotched leaves and stunted plants.

For two years the Atikokan dispute raged, even to the unusual point of engendering terse diplomatic protests from Washington to Ottawa. By March 1979 the US Environmental Protection Agency staff in Duluth published a report which claimed that Atikokan would produce enough added air pollution to tip some already acid-sensitive lakes in the region towards acid death. But larger events overtook the issue. In early 1979 Ontario Hydro suddenly slashed the size of the Atikokan project in half. The move had nothing to do with acid rain; it was part of Hydro's reaction to rising public and political alarm over the giant utility's runaway expansion plan which had created a province-wide 40 percent surplus of power production capability, a $10 billion debt and escalating electricity rates. As the Ontario environment ministry noted in a scathing reply to the EPA study, there would only be two 200 megawatt generators at Atikokan, burning low-sulphur coal imported from Saskatchewan and producing less than 30 tons of sulphur dioxide per day. There was no confirmed evidence that sulphur dioxide or acid rain from such small emissions would harm local or regional environments. And furthermore, at least 70 per cent of the region's predominant weather patterns carried winds from Atikokan east across Ontario, not south into Minnesota. When the Steep Rock mine near Atikokan closed down for good, it would cease to emit its even larger column of sulphur dioxide from the processing plant. The Atikokan power plant did not need and would not have air pollution scrubbers built in, Ontario declared. The United States should look to its own gross emissions of sulphur dioxide, including those from 22 uncontrolled coal-fired plants in southern Minnesota, if it was worried about air pollution.

It was a stinging rebuttal, one which might have prompted even further detailed American studies, more angry environmentalists and further diplomatic exchanges. (The EPA did manage to fire off, in quick reply, a note that the emissions would produce the same predicted damage over a longer period of time.) But by then acid rain had become an issue reaching far beyond Atikokan. Central Ontario's cottage-country lakes had been discovered to be acid-damaged, and crop and vegetation damage was suspected. Atikokan faded quickly as the same scientists who had been slinging studies and counter-studies at each other were reluctantly drawn together over measurements of critical acidity from Muskoka to Halifax to Florida. The politicians began scrambling for new accusations and rebuttals to what was declared a "national emergency" covering the eastern half of the continent. Ontario Hydro resumed its construction of the forgotten little power plant, without pollution scrubbers.

However, Atikokan should not be so easily forgotten. The dispute that raged there is essentially the same one which surrounds acid rain today across North America. Two neighboring countries are destroying their own and each other's environment, and neither is prepared to go first in turning off the pollution. Atikokan included intense US pressure on Canada to stifle a source of air pollution, almost minuscule compared to US emissions sweeping into Canada. It involved adamant Canadian refusal to curb that source, while it and other Canadian sources threatened to play havoc with the Canadian environment, and blow south to the US.[1]

It's ironic, or perhaps just narrow-minded, that Ontario Hydro, Ontario, and Canada chose to harden their resistance to acid rain abatement around a source that may not even be needed. As Hydro's long-range power demand projections have continued to decline, it has become clear that Atikokan is not critical to meeting future requirements in the north. The decision to cut back the plant by 50 per cent in 1979 was the first confirmation of its surplus status. And as a royal commission investigating Hydro's long-range plans concluded in early 1980, there are millions of surplus kilowatts in the neighboring province of Manitoba 200 miles west of Atikokan.[2] Ontario

Hydro, a creature of the Ontario government and long-time tool for stimulating "growth," is rushing to build Atikokan beyond a point of easy cancellation. However, the plant will provide less than 200 full-time jobs when it is completed, if it is completed. It will also produce a little more acid rain.

On January 13, 1978, the Ontario energy ministry emerged from meetings in Washington to announce that it rejected any claims that the Atikokan power plant required air pollution scrubbers, or a further study of its regional environmental risks. "It is not Ontario's intention to attend further international meetings on the subject of the Atikokan project," said then provincial energy minister James Taylor, responsible for Hydro, speaking on behalf of the Ontario government and in effect staking out the federal Canadian position. It was the first time acid rain had reached the level of a political consideration and Ontario dismissed it. Within a year the same government had become confronted by an acid rain problem which could neither be dismissed and ignored, nor apparently controlled: Inco.

The 12th largest company in Canada with $4 billion in assets stretching from Canada to Wales to Guatemala to Indonesia, Inco Ltd. (formerly International Nickel Co.) has dominated the world nickel markets and the Ontario government with powers likened to that of a king. It used to be called King Nickel. Others call it "arrogant."[3] Since 1902 the company has based its empire on the production of nickel ore which has been essential for everything from steel alloy to armor plating and guns (for both sides during World War One) to dairy equipment to elements in electric blankets and even to San Francisco's Golden Gate Bridge. At one time Inco produced 90 per cent of the world's supply of the wonder metal, and it remains the world's largest single producer today. Its sales have exceeded $2 billion in recent years. A friend and benefactor of at least one Canadian Prime Minister, Inco has also owned its own union for a time, whole towns, and the fortunes and lives of tens of thousands of workers around the world. The headquarters of Inco are in Toronto and New York;

the guts of the company are in Sudbury, Ontario where mile-deep mines and world's-largest smelters produce the millions of pounds of nickel per year.

Inco has left an indelible mark on Sudbury—some call it scars and bruises. No city in Canada is more known for its one-industry dependence than Sudbury on Inco. To many Canadians the words are synonymous. More than 20 per cent of all employment for the 85,000 residents is directly due to Inco; the company pumps more than $3 million into the local economy each week. It has also taken an average of five or eight workers' lives per year (depending on company or union statistics) in the last decade in on-the-job accidents. Working conditions for the 12,000 United Steelworkers at Inco have always been a major issue in the repeated strikes which have hit the company since the 1960s. After Inco's and the Ontario government's records of safety were scrutinized by the 1974 Royal Commission on Health and Safety in mines, Inco was ranked as worst for job safety among all big metal producers. The company and the provincial government, largely pushed by the union, are only now beginning a study of the eventual health and fate of the tens of thousands of employees who have worked in the mine and the smelter dust, heat and fumes described as "a very dirty hell." And when times are tough for Inco, Sudbury suffers first—weak nickel markets and corporate profits (due to over-production and faulty corporate expectations) suddenly threw 2,000 workers out of work in 1977. Just before the workers went out on a 8½ month strike in 1978-1979 over better wages and working conditions, they coined the slogan "Inco Is a Four-Letter Word." Burdened with immense stockpiles of nickel built up during the over-production, Inco profitably weathered the strike by selling off the stockpile. The workers weathered the winter on a weekly strike pay of $25 per single man.

Inco's mark extends for miles beyond Sudbury. In fact, the barren landscape so resembles the moon that it has been used for American astronaut training. Before 1900 the forest which surrounded Sudbury had already been thinned and reduced by massive logging and inevitable forest fires. By the turn of the century and the discovery of nickel underground, much of the remaining forest was ravaged for mine shaft construction and for fueling the fires under open piles of ore to "roast" it for later smelting. The roasting gave off dense clouds of sulphur-laden smoke which killed trees, shrubs and even grass. .The land eroded away, exposing bare, soon-blackened rock. After smelters were erected in 1930, the worst of the sulphur fumes began to spew out of stacks, and mountains of useless rejected rock tailings and slag began to pile up, stretching miles from the smelters. The smoke drifted across Sudbury, the dust and grit from the tailings obscuring the sun on windy days. It wasn't until 1958 that the company began seeding a special grass to hold down the dust on the tailings. The remaining hundred square miles around Sudbury remained a dusty, barren moonscape. Throughout the 1960s more than 6,000 tons of sulphur dioxide poured out of Inco's smelter stacks and wafted across the land each day—the largest single source on the continent.

By 1970 the yellow-brown skies over Sudbury, the biting fumes which rolled across the city on bad days, and the recurrent violation of ambient air standards prompted the Ontario government to press some controls. In July energy and resources minister George Kerr announced Inco would be required to reduce emissions from its main smelter to 5,200 per day immediately, to 3,600 tons by the end of 1976, and 750 tons by the first of 1979. The order included a requirement for a 1,200 foot Superstack to be built, "to dilute and thus disperse the smelter's gases, as it was agreed by experts that this was the quickest *remedy* for sulphur dioxide."[4] In fact, however, the Superstack scheme, like almost all of the control order details, had been "negotiated" by the company and the provincial environment ministry in closed meetings before the order was imposed. The stack idea was also Inco's—almost unnoticed, 15 months earlier, Inco had already announced plans to build it.[5] The initial impact of the order was negligible, since Inco at that point was already operating at the ordered limit of emissions and was already committed to other relatively small pollution control efforts at other sites in the smelter complex. Company officials indicated at the time they were unsure how they could cut the main smelter emissions 90 per cent within eight years, but no-one suggested it could not be achieved—at least not in any public statement.

By 1973, three years ahead of schedule, Inco had cut emissions to 3,600 tons per day, largely through construction of a new mill to separate out the sulphur from the nickel ore before it went into the smelter and up the stack as sulphur dioxide. The new mill both improved nickel production and reduced air pollution. It was Inco's finest achievement—it was also Inco's last environmental one. No further significant abatement was accomplished after 1973. And on July 27, 1978 the provincial environment minister announced that a new control order would be imposed, extending Inco's emissions for four more years at 3,600 tons per day—a limit the company had reached five years earlier—and in effect giving the company 10 years to get below 3,600 tons per day. In a single-paragraph statement, Inco termed the new order "a practical approach to a complex situation. It recognizes that significant improvements were achieved ... it directs the company to conduct studies which will assist the ministry to determine whether further action should be taken." And in what could only be taken as a threat of lay-offs if tougher controls were required, the statement concluded, "In the interim the order permits Inco to continue the operation of its Sudbury facilities." The environment ministry's explanation for the order was much longer but said the same thing. Inco had produced considerable initial abatement, there was no confirmation that any abatement below 3,600 tons was necessary—Sudbury's local air was noticably cleaner thanks to the Superstack—and the 750 ton target was "unjustifiable from an environmental perspective and unrealistic from a technical point of view."

It was a stunning decision, unexpected and unacceptable to both Opposition politicians and a vocal segment of the public. It prompted vehement criticism of both the company and the government, sparking an unprecedented coalition of 1.1 million Ontario cottagers and environmentalists who sent representatives to demand the resignation of then environment minister George McCague. Acid rain had surfaced in media reports only months earlier, coupled with the first environment ministry reports on acid-dead lakes in the Muskoka-Killarney areas, and Inco's emissions, the widening area of acid-damaged lakes, and the government's apparent impotence could not be dismissed as Atikokan had been. Within two months environment minister McCague had been suddenly shuffled from his position into relative obscurity in a minor ministry, a sacrifice to unanticipated public outcry. Another environment minister, Harry Parrott, was produced—the fourth in five years. But it wasn't until February, 1979, when the Legislature's Resources Development Committee opened special hearings into acid rain and Inco, that the government's true role became clear.

The Legislature committee's 18 lengthy sessions unravelling the acid rain and Inco affair offered a scenario of government and industry in bed together. While the winter storms raged outside, the stuffy hearing room remained packed with spectators; a dozen or more Inco and environment ministry officials maintained a permanent observer corps clustered together in the rear of the room. There was unusually consistent attendance by more than the bare quorum of legislators needed, and almost non-stop media reporting. It was the Ontario and Canadian public's most concentrated crash course in acid rain, government decision-making and corporate response. By the time the committee had heard from government and non-government environmental scientists, the long-ignored and widespread impacts of acid rain were well established, and focus turned to the bureaucrats and politicians responsible for turning off the rain. By the time the gentle inquisition was over, the government's inability to budge King Nickel and the need to retreat behind vague statistics and undefined policies were painfully obvious. Out of it came a provincial posture protecting Inco from further abatement which still influences any action anywhere on the continent against acid rain.

In testimony and documents produced before the committee, it was revealed that as far back as 1970 Inco had privately warned the government that the big reduction to 750 tons of sulphur dioxide per day was, from Inco's position, impossible. Yet Inco had agreed to the control order without public protest or appeal. The cut-back to 3,600 tons was comparatively easy; beneficial to the nickel production process too, and well underway by 1971. But beyond that, nothing could be done. By the mid 1970s Inco had, to its credit, spent $14 million studying a different technology of smelting—already in use elsewhere—which would have completely separated out the sulphur. But it was declared an unacceptable solution for "technical and economic reasons." Inco also warned it didn't know what to do with the left-over sulphur or acid. The company also tinkered with a $9 million experiment to better capture sulphur dioxide as it rose from the smelters, but that too was a failure, the company said. Environment ministry officials were regularly briefed by Inco on its failures, and by 1975 they "were getting a strong feeling" that insistence on the 750 ton limit was not realistic, according to Erv McIntyre, the chief ministry representative on the scene in Sudbury. McIntyre and other officials testified that they had no idea why 750 tons was picked in the first place, back in 1970, but Inco said it was impossible, and the ministry came to believe Inco. "This feeling was apparently based on a general perception of technical and economic difficulties rather than a thorough and comprehensive analysis of the abatement options and their costs," the committee noted after McIntyre's testimony. And in an interview months earlier McIntyre had admitted as much: "We didn't go over their economic analysis of the cost of cut-backs. They talked in generalities and we had to agree with them. The experts in this country are in the mining business, like Inco."[6]

Inco's experts dismissed air pollution scrubbers like those

applied to coal plants and smaller smelters elsewhere. Inco said it would cost billions of dollars to rebuilt a new and clean smelter. Who could question the experts? In 1975 Inco had reported to the government that it appeared possible to cut the emissions from 3,600 to 1,500 by installing more efficient furnaces. It would cost nearly $200 million, and it was only a plan, not a commitment, Inco added. It would have meant giving Inco an extra year beyond the 1979 deadline, and would have produced emissions still twice as high as the deadline target. The ministry rejected the plan, urging Inco instead to try harder to provide a better solution; Inco dropped the plan quickly, arguing costs would be an unacceptable $100 million higher than the first projected.

There was one other option, Inco suggested. It could cut emissions as required by simply cutting production of nickel. Unfortunately, this would mean closing 7 of the 10 mines in Sudbury, most of the smelter, a refinery in Port Colborne and other facilities, and dropping at least 6,500 workers. Inco would be forced to abandon Sudbury. "Obviously the consequences of this would be contrary to Inco's interest," the company added. "We believe they would also be undesirable from a public point of view." It was a warning. When the 1979 deadline rolled close, the Ontario government backed off, giving Inco four more years to suggest to the government what should be done.

Cost to Inco was the focal point of the private meetings over the years. The company argued it could not afford $300 million for even a partial reduction below 3,600 tons per day, let alone "billions" for a rebuilt smelter. These expenditures returned nothing to Inco except clean air. No return on investment, no profit. Mountains of raw sulphur or lakes of sulphuric acid would result which could neither be profitably sold nor easily stored, Inco claimed. The company preferred to invest its money on a $270 million acquisition of an American battery plant, or (bolstered with $70 million in federal loans) commit itself to $1.1 billion mineral developments in Guatemala and Indonesia in the early 1970s.

Inco's profits had sagged briefly during the early 1970s amidst unpredicted poor markets, and the company had deferred more than $200 million in taxes by 1973. But by 1974 profit was nearing the $300 million mark again, partially aided by slashing costs—and 6,000 employees—since 1972. By 1978, when the controversial 750 ton pollution target was due, Inco had accumulated $1.4 billion profits since the time the target was first imposed (and declared impossible). Lacking any detailed knowledge of abatement technology, of Inco's alleged costs, of future environmental effects or even of the company's exact financial status, the Ontario government believed everything it was told. And of course none of this was discussed publicly, although the environment minister and cabinet were kept briefed; in fact until 1977 acid rain was not discussed aloud by the ministry.

As ministry officials testified, they were quite satisfied with Inco's 3,600 ton achievement anyway, and the Superstack had markedly helped local Sudbury air, and nobody really could be sure what 3,600 tons per day meant to the atmosphere elsewhere. "We know that at 3,600 tons they [Inco] don't really contribute to an acid rain problem," McIntyre told disbelieving members of the Committee. (He later amended this to "contribute significantly.") The government's environmental experts argued that the Superstack emissions got so diluted as they blew beyond Sudbury that it was impossible to pin the blame on Inco. "I am convinced that the background [of total sulphur dioxide over Ontario] is so high in Ontario that if we eliminated all our discharges to zero ... we would still have the acid problem that we have today," McIntyre added. It was a bold claim, implying that more Inco abatement efforts were entirely useless.

The committee heard testimony from non-governmental scientists about the amounts of heavy metals from the Superstack plume which *still* fell in the Sudbury region—the ministry men admitted such levels were high enough to be toxic to fish (but not to people). The ministry also displayed a complete lack of knowledge about how much of Inco's airborne sulphur dioxide fell as dry sulphates or potentially acidic material, and about where it fell. Not knowing exactly how much acid rain could be blamed on Inco amidst other continental sources, nor detailed costs, the government felt nothing should be done, in 1975, 1978 and perhaps 1983. The ministry couldn't "justify it." Coming from the government's agency designated to protect the environment, at a time when acid rain was rapidly becoming documented as a threat of unparalleled proportion which required urgent action at every source, it was an amazing attitude.

In its concluding statement the Legislature Committee stressed its concern with the lack of ministry data on Inco and it noted that Inco would inevitably have to abate further as part of a continental control on acid rain. The committee admitted that no one knew how far Inco might have to go to reduce emissions as part of the continental clean-up. But "on the other hand, any further delay in abating emissions from the continent's largest sulphur dioxide source and any indication of a lack of commitment to solving the very serious problem of acid rain would be extremely undesirable." The committee recommended that within one year the environment ministry find out for itself, independent of Inco, if technology existed to further cut the 3,600 tons-per-day emissions to 750 tons, and if so, should order it to be in place by 1985. Costs were not mentioned as a constraining factor. And if such technology didn't exist, then Inco should be ordered to do what it had admitted it could do—cut the 3,600 tons emissions in half. This should be done by 1983. Someone had to start turning off the acid rain somewhere. The biggest source on the continent, a financially healthy company, seemed the appropriate pioneer. In June of 1979 the committee's findings were issued. The eight Conservative government members of the 16-member committee dissented from the recommendations, although they offered none of their own. Inco offered no comment on the report. The Legislature eventually received the report, as advice. And nothing happened. To 1980 Inco's Superstack and 3,600 tons-per-day limit on sulphur dioxide remained unchanged, as did the fall of acid rain everywhere.

Despite the Legislature committee's probings and revelations, an important fact was neglected, and the issues it raised have yet to be taken up. According to several disheartened federal officials, a solution to Inco's pollution emissions and the basis for a Canada-US breakthrough on acid rain had been mapped out, but kept buried, since 1977.[7] The key to the solution lay buried in Cargill Township, a deserted wilderness 23 miles southwest of Kapuskasing in northern Ontario. Just below the surface there lies Ontario's biggest deposit of phosphate rock—more than 62.5 million tons according to Ontario government reports done in the early 1970s. The phosphate is suitable for making commercial fertilizer by adding one essential ingredient—sulphuric acid. One hundred and five miles south of the untapped Cargill phosphates is an uncontrolled source of immense volumes of sulphuric acid: Inco's Superstack in Sudbury.

As far back as 1930 Inco had used limited systems to capture some of the sulphur dioxide fumes and convert them into sulphuric acid which was made available to CIL Ltd. for marketing. By 1967 the Inco-CIL sulphuric acid business involved a 1,400 tons per day operation, then the largest in the world, and CIL was dispatching entire tank trains of the acid across North America for use in everything from rubber manufacturing to medicine. Inco was disposing of its air pollution to CIL as a saleable commodity. It was done at minimal profit, and in the early 1970s CIL couldn't find sufficient new markets to make additional acid production worthwhile (and Inco scrapped a $20 million, 2,300 ton per day plant) but the process, as far as it went, worked. But as Inco argued to the Ontario environment ministry in 1975, capturing more sulphuric acid would require very expensive reconstruction of the Sudbury smelter, and worse, there was no market for the acid. During the Legislature committee hearings Inco officials emphasized their problem of huge reserves of unsellable acid, if the company was forced to spend millions (then escalated to $1 billion) to sharply reduce emissions. And as late as November 6, 1979 Ontario environment minister Harry Parrott warned the Legislature of the spectre of lakes of sulphuric acid forming around Sudbury. "Those who demand removal of acid from Inco's Superstack emissions are being simplistic. The problems have yet to be worked out." And Inco vice-president Stuart Warner repeated: "We already have too much acid. The market is limited."

But at the Cargill Township phosphate deposits the owner firm was complaining it couldn't find affordable sulphuric acid in large enough supply to make processing the phosphates into fertilizer worthwhile. "We've talked to all the acid producers, including Inco," Frank Piper, corporate secretary of Sherritt-Gordon Ltd. said in an interview in late 1979. To produce fertilizer at the remote site at world-competitive prices, the company needed low-priced, "almost free" sulphuric acid, and it hadn't been able to find any. The Cargill phosphate develop-oment was in limbo, he said. And yet, in another interview, Inco vice-president Warner, while discussing eventual Inco abatement technology, had offered: "We'll probably have to do it by producing sulphuric acid. We'll have so much acid we may have to give it away." (Inco faced no deadlines for substantial abatement at that time.) Warner also argued that there was no market for sulphuric acid. "Most of it is used for phosphate fertilizer. Most of the phosphates are in the US or South Africa." He didn't mention Cargill Township and the 62 million tons of phosphate, 105 miles north of Sudbury.

And yet, in Ottawa, officials of the federal Energy, Mines and Resources ministry had repeatedly suggested linking up Inco's acid potential and the Cargill phosphates. Their reports, like the phosphates, had been kept buried, although they contained valuable resources for forming an abatement policy for Canada. Led by Gary Pearse, then the ministry's sulphur commodities expert, they had pointed out that the world demand for sulphur products was drastically changing. Certainly in 1975 and 1976 sulphur itself had glutted the market, selling for as little as $5 per ton at the western Canada natural gas plants where it was automatically scrubbed from the gas and stockpiled on the prairies. But by 1977 a demand increase was predictable. With impressive foresight and awareness of acid rain, Pearse's research team proposed Canada and the US set a joint agreement: sulphur and its acid for all industrial uses in eastern North America should be produced only as a byproduct of air pollution controls, captured from smelters and even coal-fired power plants. Western natural gas sources and the traditional Louisiana raw sulphur-mining sources should be channeled abroad to meet increasing world demand. It would require continental fixing of markets and distribution boundaries, but it was not impossible, with sufficient political initiative.[8]

The response to the National Sulphur Marketing Strategy proposal was "heartbreaking," said Pearse in an interview in late 1979. Another member of the team said "External Affairs was sent our proposal as a possible negotiating stance to use in meeting the Americans and they rejected it. And the federal environment ministry was an even bigger roadblock. Their lack of interest was simply astounding." That was in 1977, when Canada was first dickering with Washington over the Atikokan power plant emissions, during the tenure of Len Marchand, the federal environment minister who ignored acid rain. Nonetheless, the sulphur strategists circulated their report widely in Ottawa, and to the Ontario government, and even Inco acquired a copy, said Pearse. It was never discussed publicly. Two years later, in late 1979, sulphur was selling at seven times the 1975 price at the Alberta gas wells which were already producing less and less sulphur from "cleaner" gas and slowly reducing their stockpiles. On the world spot market sulphur had hit $105 per ton. Iran was cut off as a major world supplier, the Soviet Union was locking down long-term, high-demand contracts for internal use, and the Louisiana sulphur mines were slowly folding under the impossible costs of energy required for the mining. A June, 1979 federal government study predicted skyrocketing demands for sulphur, acid and sulphur-phosphate fertilizer.[9] Food-hungry Asian, African and South American countries couldn't get enough fertilizer and would be desperately short within 5 years, according to an updated version of the Strategy in 1979. "Canada could sell another 3 million tons of sulphur abroad right now. Within two years there will be a world shortage of sulphur for fertilizer," said Pearse in late 1979.

Pearse said he couldn't pin down exactly why the Strategy reports had been kept buried in 1977, unless it was due to the reports' analysis which concluded that the federal government might have to take bold steps in market and price regulation instead of just threatening clean-air regulations. And it required US agreement. Another member of the study team said Inco had turned down the idea because "Inco sees acid production only as an unprofitable investment, rather than the cost of pollution abatement. It's cheaper for Inco to do nothing and deny responsibility for acid rain as long as possible." A third member noting the strategy suggested that government loans to Inco might be needed to spur the company into reducing emissions and producing the essential byproduct for a booming Canadian fertilizer industry and world sulphur market added, "It might not be profitable from the narrow corporate point of view but if we're serious about the acid rain our strategy was the least expensive Canadian solution."

Federal environment minister John Fraser said in September 1979 he had never heard of the Strategy studies. Inco vice-president Stuart Warner denied any knowledge of them. And federal environment ministry officials said the study had never been circulated in that department. Early in 1980, after details of the Strategy appeared in *The Toronto Star*, federal officials conceded it was being dusted off, to be revised by omitting the recommendation for government loans to Inco. "It's being brought to the front burner now. Sulphur marketing probably is central to the issue of acid rain control." Within weeks it was released, unchanged, unrevised. For Gary Pearse it hardly mattered. He'd quit weeks earlier. "The morale around here is very poor. There are too many vested interests." The other key members of the strategy team had already transferred out of the department.

In April 1980 new federal environment minister John

Roberts began referring to "new" federal studies which he said proved Inco could reduce its emissions 50 percent or more. One of those studies included a description of technology to produce sulphuric acid for phosphate fertilizers—using "local" phosphates.[10] The local phosphates are buried underground, in Cargill Township. They will almost certainly be part of the solution to safely disposing of Inco's acid rain-causing emissions, whenever that happens. Inco knows when. Inco has a plan to develop a new smelting process capable of producing much cleaner emissions and much sulphuric acid.[11] It will be very expensive, costing perhaps $1 billion, and will require government aid in carving out new markets for the acid, says Inco. It will take many years to perfect and install. In the face of no tough deadlines for pollution abatement, only Inco knows how many years.

# 8
# The Ottawa Connection

The Americans threw the first punch. It was a US Senate resolution in May, 1978, complaining about Atikokan emissions which raised acid rain to an international issue. The US was protesting a single but very specific Ontario source, but Ontario—and the federal Canadian government—had no defined list of power plants on the American side to complain about. Details of the transboundary exchange of sulphur dioxide was largely uncharted then, but the Atikokan attack put Ontario on the defensive. Frustrated and feeling abused, the province pulled out of international discussions on Atikokan to rethink its argument, although still adamantly determined to do nothing about anything.

The pollution extension for Inco swept Atikokan into the background by July 1978. But even in its first press release in justification of the Inco order the Ontario environment ministry began to outline a policy it would rigidly hold to for almost two years. "The transboundary movement of sulphur dioxide ... is the major factor in the acid precipitation phenomenon. ... [It] is global in scope ... The complexity of tracking air masses that travel hundreds of miles from one area and one country to the next underscores the need for more study of this serious problem,"[1] the release said. The ministry admitted it was in poor shape to document how much acid rain was falling where, let alone from what source. (At least partially because for two years prior to 1978 it had committed almost all of its resources to studies on Sudbury's Superstack.) The need for more analysis before policing the sources was to become the cornerstone of Ontario's position. Dr. Stuart Warner, vice-president of Inco, underlined the real meaning of this do-nothing-hasty policy when he told the Legislature Committee, "We believe it would be inadvisable to make important regulatory decisions before [detailed scientific assessment] is available. Doing so could result in a net loss for Ontario." In other words, pushing Inco too hard too soon could cripple Sudbury.

Sudbury regional environment chief McIntyre told the committee that forcing *all* Ontario sulphur dioxide sources to halt emissions would bring no relief from acid rain, unless major reductions were imposed in the US too. Again, he had no data to prove his claim, but an old Atikokan argument was being revived: the US was to blame. But a more sophisticated phrasing had taken shape since the Atikokan debacle. What was needed, Ontario and federal officials testified, was an international agreement by Canada and the US jointly to take action against the problem, possibly based on a step-by-step timetable of emission abatements on both sides of the border. The international agreement on cleaning up the Great Lakes was offered as an example. An international treaty was already in very preliminary discussion, thanks to the initial diplomatic protest raised by the US over Atikokan. Ontario and Canada still needed precise data on where American acid rain was falling on Canada, and at what cost. Environment Canada's Ontario regional director Bob Slater told the committee that "the treaty could be ready within a year, and the specific regulations to implement it within three years." In other words, by 1982 continental acid rain would start being controlled. The Ontario environment ministry officials nodded in agreement.

Ontario's original acid rain researcher Harold Harvey quickly pointed out that given both Ontario's and Ottawa's poorly-funded and poorly-motivated acid rain research programs, five years or more was a likely timetable for a treaty. Dr. Jim Kramer of McMaster University suggested it would help if Ontario and Ottawa developed some sort of team approach to chasing down the needed facts. Members of the Legislature Committee wondered aloud if Ontario should not show its good faith by first doing something significant, such as putting tighter controls on Inco. William Glenn, a researcher from the Pollution Probe citizens group, warned that the Atikokan impasse would loom again in the international negotiations. Regardless of the gross emissions from its existing source, the US would protest that it had already put tough controls on all new sources, while Ontario had backed off on Inco.

Nonetheless, the Ontario and federal Canadian position had been staked out: abatement at home as soon as the Americans could be negotiated into a similar policy. It seemed eminently fair between good-neighbor countries with a long tradition of co-operation. The realities of that tradition, when it came to politics and money for environmentalism, were barely referred to. The Great Lakes clean-up, for example, had taken 40 years. Canada had repeatedly taken the initiatives and waited for the US to follow on the Great Lakes. Within days after the Legislature Committee hearings, Ontario began refining its message. In a speech to the annual conference of Anglers and Hunters in London, Ont., on February 24, deputy environment minister Graham Scott raised all the important points. The ministry's data showed that 80 per cent of the acid rain affecting Ontario came from the US. The ministry was diligently researching the impacts. But "while research is an essential component of our action plan it must also be linked with positive international action," Scott said. He listed the major steps "which the Ontario government wants taken, which include: abatement on an international scale, preventative and remedial action, and continuing scientific investigation—our most urgent and vital need." Scott mentioned lake liming as a "remedial action," without mentioning its dubious success rate; and quoted a letter from Ontario Premier William Davis to Prime Minister Pierre Trudeau "emphasizing the urgent need for the resolution of the problem" and urging Canada to seek an international agreement and program as quickly as possible.

So the ball was passed to Ottawa's court. In May, a new Conservative federal government squeaked into power in Ottawa and federal environment minister John Fraser began staking a claim to fame on acid rain, much in contrast to his Liberal predecessor Len Marchand who ignored the issue. Ultimately, it would be Fraser's responsibility to drag the tired old Ontario "policy" into the treaty negotiations. As a federal minister only he could directly deal with foreigners; it would remain up to the provinces to make their own commitments to Ottawa, although Fraser declined to make public what commitments the provinces, particularly Ontario, had promised. That same month the US Environmental Protection Agency unveiled its final, mandatory regulations on new sources of sulphur-dioxide air pollution, with considerable publicity and impressive statistics to justify the costs and benefits. But the existing sources and what should be done about them were given no mention in the voluminous official statement, press release, fact sheets and background papers. American environmentalist groups launched immediate attacks on the serious compromises which had been made in drafting the final new-source standards, but in Ontario the stand-up speech with its optimistic Canada-US tandem attack on acid rain continued in vogue at Rotary Club luncheons and cottager association meetings.

On July 9 the International Joint Commission, which for years had been best known for its exhortations over the slow progress in cleaning up water pollution in the Great Lakes, took to the air. Acid rain was damaging the ecosystem of the entire Great Lakes Basin, the IJC's Science Advisory Board warned. Everywhere in the Basin (from Montreal to Minnesota and Chicago to northwestern Ontario) rainfall measured 5 to 40 times more acidic than normal rain. The board itemized acid-dead lakes, sterilized soils, damaged vegetation, possibly toxic levels of heavy metals in drinking water as confirmed effects of acid rain; cited US figures attributing $2 billion in architectural and $1.7 billion in health costs to sulphur dioxide; and, not surprisingly, recommended more study. It also noted neither Canada nor the US had any existing laws or treaties which would effectively require both countries to clean up the acid rain sources. "Based on past performance"—the Great Lakes clean-up—"a decade may pass" before any real pollution abatement would take effect, the IJC concluded, bleakly.

One week later, US President Jimmy Carter announced his $142 billion plan to save the nation from its energy crisis. The plan called for conversion of oil-fired power plants to coal, a doubling of the existing coal-fired power plants, and expansion of coal plants into the previously clean-air western states. Even with the controls on the new plants, the energy onslaught was going to increase the US total emissions of acid rain, numbed environmentalists and administrators in Washington admitted soon after Carter stopped speaking. "Yes, it may mean that we won't solve the acid rain problem as quickly or as best we can," one admitted. "I'm sorry," he told this author.[2] Existing sources were not mentioned.

In Ontario, environment minister Parrott termed the Carter plan "devastating." It would bring "massive new floods of acid rain to Canada" if it was pursued without tight controls on new coal-burning power plants. Ontario would step up its pressure for an international agreement on air pollution abatement, Parrott promised, and his federal counter-part John Fraser immediately concurred. Parrott did add that "Ontario industries are not blameless and we are taking steps to improve their performance," although Inco, he said, was to blame for only eight to 10 per cent of Ontario's acid rain. Nonetheless he would be meeting with Inco within 10 days to discuss "tough but reasonable" new deadlines for Inco.[3]

Nine days later, on July 26, the pressure for an international abatement treaty suddenly appeared to have produced results. External Affairs minister Flora MacDonald announced Canada and the US had reached some agreement on the subject, and a joint statement was issued. But a careful reading of the statement showed it was simply an agreement to agree on the need for a treaty, and a promise to hold more meetings. The statement listed seven principles as a possible basis for an eventual treaty, including more research, improved exchange of scientific information and complementary legislation. Ironically, one of the most important principles—"expanded notification and consultation on matters involving a risk or potential risk of transboundary air pollution"—had already been jeopardized by Carter's go-coal energy speech. Nobody had told, let alone asked Canada about the plan and its impacts, before it was announced.

July rolled into August. Federal environment minister John Fraser flew back to Ottawa from meetings in Washington and Toronto to declare that the Americans had agreed on the need for an air pollution treaty; that Ontario had agreed it would impose whatever pollution controls necessary to meet any such treaty. Fraser also had promised the Americans he would force other Canadian provinces to impose similar controls. To clear up any misinterpretation, Ontario's Harry Parrott dashed off a letter to newspapers stressing that "Ontario is prepared to take action in conjunction with the rest of Canada and the individual American states."[4] He said nothing about taking initial actions, emphasizing the "majority of acid rain occurring in Haliburton/Muskoka has its source in the south," and said it was impossible to tell where Ontario's acid rain was going. The do-nothing policy remained intact.

Meanwhile in Washington, both Ontario and Canada were being lumbered for doing nothing. "There's no question we're putting more acid into Canada than vice-versa," said Konrad Kleveno, chief of the Environmental Protection Agency's Canada-US relations branch, in an interview.[5] "But there is also the matter of Canada getting its own house in order. There has to be a reciprocal action on the Canadian side, if anyone expects the US to clean up its sources. So far we've seen no real action north of the border."

On October 15 the first joint Canada-US scientific assessment of acid rain was released, with its documentation of acid damage to lakes, its confirmation that forest and crop damage was possible, references to property damage, and hints of alarm over toxic metals in drinking water. But the report also carried the conclusion that roughly half of the total sulphur fall-out in Canada came from Canadian sources. One particularly troublesome part of the study (an American contribution) calculated that in southern Ontario and Quebec nearly twice as much native sulphur compounds—an average of 105,000 tons per month—were falling there as came in on winds from the US. Undaunted, Parrott told the Ontario Legislature one day later it would be "unfair" to curtail those Canadian sources before an international agreement was reached. Questioned about Inco, he replied "you can't say to an industry 'You alone must tackle the question.' "[6] Six days later, Inco weighed in with its own pronouncement. "We can neither deny nor accept responsibility for acid rain .... [it] happens from such a variety of sources it's almost impossible to attribute damage to a single source," said Dr. Stuart Warner, Inco's vice-president for environmental affairs.[7]

The release of the joint Canada-US report at last triggered a flood of attention and publicity about acid rain. Question period in the Ontario Legislature became a Chinese acid torture for Parrott, with opposition critics forever linking the government and Inco in the same sentence, and demanding to know why no action was being taken. On October 25 while being grilled (again) by the Legislature's Resources Development Committee which had labored through acid rain hearings eight months earlier, Parrott suddenly reversed his position. Ontario was "prepared to act singly and in advance of other jurisdictions" to take action against acid rain sources in Ontario. And Inco in Sudbury would be his first target.[8] It was a stunning flip-flop, which appeared to undermine almost five years of private and public defense of the claim that major abatement at Inco was neither affordable nor important, particularly without parallel American action.

"What's going on?" Liberal Opposition leader Stuart Smith demanded in the Legislature the next day, noting Parrott's apparent sudden change in policy. The Premier of Ontario, William Davis, set everyone straight. Ontario's approach to acid rain was to move on the acid rain question "in a balanced way." Acid rain was only a recently recognized problem, and sulphur dioxide might be less important than nitrogen oxides (Inco's emissions were nitrogen-oxide free), the premier said, apparently unaware that nitrogen oxides constituted only 30 per cent of the acid rain problem in North America. But the premier pressed on. If Inco was singled out for a pollution control order, it would mean layoffs of workers in the Sudbury area. "I am not prepared to have the government create an economic problem for the city of Sudbury. We can't do this in isolation." Davis emphasized Inco's previous abatement efforts—he called them substantial—and called the 1970 order for Inco to cut its emissions to 750 tons per day by 1978 (an order which was rescinded) unrealistic. "The technology doesn't exist to make it economically viable," he concluded.[9] In the Ontario cabinet there is an iron rule of solidarity—no minister speaks out in opposition to a cabinet decision, and no minister contradicts his premier. Ontario environment minister Harry Parrott said nothing.

On November 2 and 3 more than 800 people from across the continent crowded into a Toronto hotel for ASAP—Action Seminar on Acid Precipitation—the largest citizens' conference on the subject ever held. For many of them, the technical sessions on lake deaths, potential for vegetation damage, health risks and the warnings about "environmental catastrophe" only confirmed what they already knew. As conference co-chairman Donald Chant, vice-president of the University of Toronto and long-time pollution-fighter, told the delegates, it appeared citizen concern over acid rain was "far in advance over that of both countries involved." Federal environment minister John Fraser sought to assuage that impression in the first political speech of the conference. He promised the issue would remain his number one priority, and listed at length the research projects his ministry had begun. He described acid rain as far too serious to wait for an absolute scientific certainty—some action would have to be started based on best available, best educated conclusions. He had found American officials "appropriately concerned" about the problem, and actual abatement would have to be applied very soon—"within the next couple of years"—even if it cost $500 million per year for 20 years in Canada. A polished speaker, Fraser roused the conference delegates to their feet with his speech, although in the corridors afterwards delegates began pondering exactly what he had promised.

It was a hard act to follow, and Ontario environment minister Harry Parrott threw away much of his lengthy prepared speech when he addressed the conference the next day. But he kept the critical lines: "Many people suggest that Ontario must clean up its own sources first before we can expect the US to act. I do not agree. We must make and obtain a commitment to act. We must also know what sources are affecting what areas, and to what extent. Then we must devise practical abatement programs." It was an old and familiar theme, and roused no one.

In a speech which federal bureaucrats still dredge up when they're accused of dawdling, Canadian environment minister Romeo LeBlanc announced in June 1977 that Canada and the US were "creating an environmental time-bomb" with transboundary air pollution. His speech to a conference of air pollution experts in Toronto was unusually prescient for 1977. His speechwriters found it necessary to put quotation marks around "acid rain" in introducing the issue to a public forum. LeBlanc claimed that "air pollution is a political problem, of major proportions. We do not have time to wait for final research before beginning political action." But in an interview he repeatedly stressed that an agreement was necessary for a joint Canada-US crackdown on long-range air pollution. He refused to indicate Canada would act alone in cutting down its air pollution.[10] LeBlanc may have sounded the first federal warning on acid rain, but he had also cemented a federal policy on abatement which was to last almost as long as Ontario's: wait for the Americans to act first.

By February 1979 the Inco affair was being probed in

Ontario by the Resources Development Committee, and in Ottawa the federal government was pressed to explain its policy on acid rain abatement. Len Marchand, then the Liberal environment minister, reminded the Federal Parliament of earlier statements which "indicated we would be signing an air quality treaty with the United States. My officials, along with Ontario officials, have held meetings with US officials. We are moving as fast as we can."[11] It was a typical Marchand utterance, both because it said very little, and was in response to a direct question on the subject. Unlike LeBlanc, Marchand rarely spoke out on acid rain, unasked. Marchand came from the forested west coast riding of Kamloops-Caribou and had established himself as most interested in environmental management, rather than protection. During the Ontario Legislature Committee's hearings, members of the committee took Marchand's officials to task for their low profile on acid rain. "I apologize if we seem too calm. We do see it as a matter of significant urgency and national concern," Environment Canada's regional director Bob Slater replied to the committee. He stressed the need for, yes, detailed research. A US-Canada treaty could be "ready within a year, and the specific regulations to affect acid rain within three years," he assured the committee, noting, however, that it had taken 40 years to put an effective international agreement on cleaning up the Great Lakes into effect.

Harold Harvey, the University of Toronto scientist who had lost his experimental fish population to acid rain in the Killarney Lakes in 1969, and had drifted in and out of federal government jobs since then, was far less optimistic, and less charitable. "With the exception of Atmospheric Environment Service [the federal weather service monitoring acid rain], when you talk to the feds all you get is a large sucking sound," Harvey declared, and predicted it would take at least five years for any effective treaty. As Harvey explained later, the federal ministry had devoted far less research into acid rain than its own and outside scientists had recommended, for years. In Ottawa, some federal environment officials conceded they had been late in taking acid rain seriously. The environmental protection wing of the ministry had been repeatedly pared down in funds, and the ministry itself had an unclear role—it had been re-organized, amalgamated and separated four times in six years. And it couldn't order individual provinces to control air polluters even if it wished to. "We do believe in the philosophy of containment at source. High stacks are inappropriate because they only spread the problem further," Ray Robinson, assistant deputy minister, explained. "But the provinces, and the US for existing sources, prefer to legislate ambient controls."[12]

All Ottawa could really do was research the problem, and talk to the provinces and the Americans. Robinson's minister, Marchand, insisted "we've really been quite active to date. It won't take five years for a treaty. I suppose, yes, there will be some damage in Canada in the meantime but I don't want to be an alarmist. We're moving quickly."[13] But in Ottawa as in other governmental bureaucracies, politicians come and go while bureaucrats remain. Real policy usually moves only as fast as they do. And as Robinson argued, "I'm aware of the reality. The power lies not in Ottawa or Toronto but in Washington. Our only trump card lies in the fact that they too are suffering acid rain damage. If they let it go too long, they'll destroy their environment too."

In May, John Fraser had come to the environment portfolio, as part of the change to a Conservative government. Firmly wedded to a policy of wide open co-operation with the provinces, the new government appeared unlikely to attempt pressuring anyone in Canada to tackle their acid rain sources. But Fraser was an ambitious man, a one-time environmental lawyer on the west coast who had sought the party leadership, and was determined to keep a high profile handy for the inevitable next leadership race. Fraser tackled acid rain in speeches, meetings with the provinces and the Americans, and in easy accessibility to the media. By June he had rankled independent-minded Quebec. There, environmental minister Marcel Leger complained bitterly of Fraser's criticism of slow provincial action against air polluters—and Quebec had a huge one: Noranda Mines' smelter at Rouyn-Noranda.

Leger warned that "the federal authority is seeking, under the pretext of international issues, to intrude even more into fields of provincial jurisdiction, which for Quebec is unacceptable. The problem will become political, even constitutional."[14] Leger's ministry at that point was only beginning its first major research program into acid rain. Leger's complaint about federal intrusion was the first time his government had officially admitted there was any acid rain problem in the province, regardless of whose responsibility it was to turn it off. Further east in Nova Scotia, environment minister Roger Bacon was also just waking up to acid rain: "We are aware of the situation and we are somewhat concerned. We will be monitoring it on a yearly basis."[15] The important monitoring and findings of an acid-depleted salmon fishery in Nova Scotia had been done by federal officials. In late June prime minister Joe Clark squeezed in a plea for fast American action on its coal plants in the northeast, during the Tokyo economic summit. In early July the International Joint Commission concluded it could take 10 years to achieve an effective treaty.

Fraser rebounded, with a promise to speed up the international pressure to halt "this appalling situation. I'm satisfied we know enough about what's happening to go ahead with the political steps that are required. The public is not going to sit back and tolerate a government that does nothing."[16] And yet, Fraser ruled out independent action by Canada. One day later, President Carter announced his massive go-coal policy, and Fraser was appropriately worried by US intentions: "There must be no environmental sacrifices ... I can't believe any US administration today would seriously contemplate increased environmental damages. It will be politically costly to Carter if he doesn't offer assurances."[17] Or as one of Fraser's senior officials had argued, only a few months earlier, "Thank God for US environmental extremism—when the irreversible damage is calculated, all hell will break loose in the US."

By early August, Fraser was saying he had found

Washington officials "very concerned" about acid rain: "they want an agreement, we want an agreement. The question now is how do we put together the mechanisms in achieving that agreement as quickly as possible? We do have a time problem. If the emissions sources continue in fifteen to twenty years forty eight thousand lakes just in Ontario will be dead." But as Fraser also explained, the US was already driving a hard bargain. The Americans wanted a firm Canadian commitment to match both new-source standards and the American approach to existing sources: tough ambient standards, high penalties for violation, and insistence on best available control technology —none of which existed in Canada, much of which was yet to be imposed in the US. They wanted each province to meet these standards and strategies, voluntarily or by federal coercion.[18]

John Fraser was caught in a dilemma. His federal government could intrude into provincial environment jurisdictions only where specific federal legislation existed, or where Ottawa was acting in the name of enforcement of a federal commitment to a foreign power. Ottawa had no legislation at all relating to acid rain. And to get the Americans to agree to a treaty which would give him that power to intrude into the provinces, Fraser had to already have provincial agreements to take those actions. Ontario, to cite only one province, was not prepared to take any action before the US did too.

Ignoring Quebec's rumblings of independence, Fraser talked of "transferring provincial powers to the central government." Fully aware of Inco's cost claims, he hinted it might require government tax incentives and subsidies to overcome corporate resistance. But clearly the federal options were not at all clear nor mapped out. As Fraser said, "I'm prepared to look at any means whatsoever to accelerate the installation of necessary emission procedures. And in any answer I can give today I don't want to preclude any particular approach, novel or different or unconventional, that we might bring to bear. ..."[19]

Fraser's staff, who continued to meet with their American counterparts in occasional, informal sessions, relayed American complaints back to Fraser, but his response, according to him, was "don't tell me the difficulties, get on with it." In a lengthy and apparently candid interview on September 21 in Ottawa,[20] Fraser conceded the difficulties of talking the Americans into an effective treaty soon, "just as their administration is going on Hold until the presidential election is over." But he talked confidently of "an explosion of national Canadian concern such as the Americans have never had, if we have to face a decade before abatement of acid rain. The US cannot afford to have hostile Canadians for neighbors." He said he was "counting on the co-operation of Ontario and Quebec" in advancing abatement policies which would make federal interference unnecessary. And finally, he observed that "action first by Canada is one of a number of options. We've considered it. Clearly action by Canada first does much to advance our argument to the US concerning the importance of the issue."

Somewhere between that interview and a public speech to the Air Pollution Control Association in Ottawa on September 25, John Fraser ruled out some options. He praised the spirit of Canada—US co-operation. But he did not include the key lines from the written speech: "Our appeals for more US controls will likely fall on deaf ears if we do not begin to clean up our own act. American environmental laws have on the whole been tougher than ours in the past and we have to make sure that the application of the laws that are necessary and appropriate to Canada will be just as tough on our side of the border." And in a press conference following his speech Fraser spelled out his position: the federal government would not introduce strong pollution controls until it could get the US to move at the same time. "At this time my strategy is to move equally with the Americans."[21] It was admittedly a period of the Conservative Government hard-sell on its policy of federal-provincial co-operation. And Fraser had begun drawing on an advisory team of environment ministers from all Canadian provinces for both advice and commitments to action—who knows what they advised?

By October 16, Ontario environment minister Harry Parrott picked up his cue: his government was not prepared to clamp down on companies that caused acid rain air pollution. His particular reference to Inco was startlingly similar to Fraser's September 25 speech: "You can't say to an industry 'You alone must tackle the question.'"[22] And Premier William Davis guided Parrott back onto the correct path 10 days later when the provincial environment minister suddenly suggested Ontario would go it alone: "We can't do this in isolation," Davis said. Ontario was going to move against acid rain "in a balanced way" and singling out Inco would only create economic hardship for the company and its Sudbury workers.

By the time of the international citizens' conference on acid rain in Toronto in November, Fraser had refined his strategy to one of Canada first taking action against its polluters perhaps "within two years." Nothing would likely be accomplished with the US until after the presidential election. And that old favorite, more research, was needed to pin down details about acid rain source emissions, rates and locations of deposition, and potentials for control technology. By November 8, Fraser had formed a special committee of environment ministers in Ontario, Quebec, and Nova Scotia to help him negotiate the effective treaty. He predicted it would take at least a year to work out the details of what Canada wanted in a treaty. "I can't take a hard line as to a deadline for imposing controls until we've done the work we're doing in the coming months. I don't know enough yet," he said. "But there is no foot-dragging, as far as I am concerned. Ten years ago rhetoric was enough in the environmental battle and the things you were fighting were oil in the water, and dirt in the water and smoke in the air—things you could see. The battle we're now fighting and which is becoming clear to the public is insidious things you can't see like PCBs, toxic wastes and acid rain."[23] And invidious politics?

# 9
# South of the border: America First?

When US Secretary of State Cyrus Vance came to Ottawa in November, 1978, to discuss auto pacts and international economics, he also carried that Senate mandate to press Canada into discussions concerning the Atikokan, Ontario and Poplar River, Saskatchewan power plants. The Senate protest was powered by much more than Minnesota and Montana senators intent on irritating Canada for the sake of back-home publicity. In 1977 the US Congress had finally wound up debate over amendments to the sweeping provisions of the Clean Air Act first imposed in 1970. The 1977 amendments had been much more bitterly fought than the ambitious regulations of 1970, and included steps to bring into line dozens of states which still lacked attainment of required air quality.

The amendments had been intensely fought by American utilities, who complained about the cost of the technology, and by states such as Ohio whose coal mines, steel mills and power plants profited best by unimpeded mining and consumption of cheap coal in massive volumes. And states like Minnesota and Montana were designated as clean-air regions and allowed no significant increase in air pollution. Any emissions blowing south from Canada would use up their small remaining pollution "allowance." Canadian emissions carried south would effectively block these states' chances to attract more industry. Promoted by apparent environmentalist Jimmy Carter, the 1977 amendments had passed into law and firmly established an impression among legislators and interest groups that the

US indeed had set itself very tough air pollution controls. Minnesota and Montana had little trouble in 1978 in pressing through the Senate the resolution urging Canada to control its sources which appeared to jeopardize the intent of the US Clean Air Act amendments.

The 1977 amendments were indeed forward-looking, but they turned a blind eye to the largest current problem. Legislation requiring tight controls on the existing sources was not written into the 1977 amendments, and short of a national emergency such legislation would have to wait until 1982 when the Clean Air Act came up for its regular review and revision. Even more so than in Canada, at least five years of accumulating evidence on acid rain had yet to be brought to the attention of law-makers and the public. And certain interests with awesome political power among coal miners, heavy industry and privately-owned utilities were determined that no such emergency would be declared. They threatened extreme increases in the rates they would pass on to consumers if forced to control their existing sources. More importantly, they could marshall an intimidating lobby against Jimmy Carter in the 1980 election.

In the face of such internal dissension, American negotiators accompanying Cyrus Vance to Ottawa in November, 1978, could hardly fail to point out to Canada that it lacked fighting spirit on new sources of sulphur dioxide such as Atikokan. The Americans also mentioned that the federal representatives, the only Canadian politicians they could confront, had no power to impose new source controls, a provincial responsibility in Canada. In reply, the Canadians failed to note that tougher US air pollution regulations were badly needed there: at the end of 1978 at least 20 states still suffered large areas where the mandated maximum levels of sulphur dioxide were regularly exceeded, and some states were balking at the rules. In early 1979, in an effort to bargain some recalcitrant states into air clean up, the Environmental Protection Agency quietly proposed to allow an estimated 145 existing power plants to boost their stack heights as a first step towards total emission reductions. As the Natural Resources Defense Council quickly pointed out, tall stacks were visibly the cause of much of the long-range transport of air pollution, and in fact had been virtually outlawed in the early 1970s. Taller stacks now would mean further long-distance acid rain. The EPA scheme never became a formal public proposal (although it remains alive in the halls of EPA headquarters). Thus it prompted no Canadian comment, but was an early-warning signal of the vast chasm between US expressions of concern and effective action.

And yet, in Ottawa, Canadian environment minister Len Marchand assured the House of Commons in February, 1979, that discussions with the Americans for an international air quality agreement were proceeding well. His staff told the Ontario Legislature committee the same thing; the treaty could be signed in a year and effective regulations be put in place within three years. It was to become an often-repeated Canadian political response in 1979 to public inquiries about acid rain and the likelihood of an early end to it. Largely unknown to most Canadians, the claim was based on ill-founded Canadian hope, some degree of deception on both sides of the border, and a distinctive Canadian blindness to the realities of American power politics. US Congressional awareness and concern for acid rain was demonstrated in April, 1979, when an EPA request for $3 million in acid rain research funds was voted down.

Early in May the Carter administration issued its National Energy Plan II, a long-range political promise of energy availability for America. It was based on coal—doubled production and use by the year 2000. But sulphur dioxide emissions would remain constant, due to the tough new-source standards mandated in 1977 and due for final declaration soon, the Plan promised. It lied. On May 25, Environmental Protection Agency administrator Doug Costle announced those final mandatory controls on new power plants. Although tougher than existing standards, the new rules did not require all that was possible of abatement technology. Combining the uncontrolled existing emissions—75 per cent would still be operating in 1995—with the emissions from new tough-but-not-totally controlled plants would mean US sulphur dioxide pollution would *increase* another 2 million tons, to more than 20 million tons per year, by 1995. This surrender had come after weeks of what one journalist termed "hard-ball arm twisting"[1] by such senators as Robert Byrd of West Virginia and Wendell Ford of Kentucky, who wouldn't support Carter's Strategic Arms Limitation Treaty and big-oil Windfall Profits Tax if tough pollution controls were imposed on coal. Democrats Byrd and Ford came from big coal states which the American Coal Association claimed would lose up to 85 per cent of their markets if controlled. Cleaner western coal would steal the markets, the coal lobby claimed. "Doug Costle is part of the administration and he doesn't want to harm other administration objectives," a Carter official explained. Costle conceded the weaker new source standards, Carter won his support, and North America gained more acid rain. Interestingly, that same month a major study by the Congressional Office of Technology Assessment entitled *The Direct Use of Coal* concluded that even the dirtiest eastern coal was usable in virtually fail-safe scrubber systems under very strict controls. It would increase the cost of burning coal, but it worked. No state would have to lose coal production if the utilities were prepared to apply the technology, to bear and to share the cost. The report came too late, and gained little publicity. Hard-ball arm twisting had already prevailed.

On July 10 the International Joint Commission released its annual report, warning that acid rain would intensify unless "swift, decisive and widespread action" was taken immediately. The major impediment to rapid implementation of control measures was not technical but "economic and political," the IJC added. In Ottawa John Fraser promised "fast action" by his government against acid rain, including early achievement of a Canada-US air pollution agreement. On July 15 Jimmy Carter made it clear that his primary economic and political priority was energy, in his message to the nation delivered after days of seclusion at Camp David. He called for a ceiling on oil imports. Oil use for electrical power generation would be cut by 50 per cent, and coal would be the replacement fuel, as well as a source for synthetic fuels in an $88 billion syn-fuel crash program. The $142 billion total "war on energy", which included energy conservation, mass transit funding and special oil profit taxes, would require some sacrifice on the part of Americans, but it could and would be done, Carter promised.

The details of how it would be done remained vague, but environmentalists, the US Environmental Protection Agency and Canada took little time to work out details of the sacrifice Carter had in mind.[2] The conversion of oil-fired power plants to coal, the expansion of coal-fired plants beyond even the National Energy Plan II of May, 1979, the stripping and mining of coal from both the eastern and western coal states, the squeezing of oil from shale rocks in the far west, even the synthetic fuel plants processing vegetation for gas and alcohol, presaged an immeasurable increase in sulphur dioxide (and carbon dioxide) emissions in the US. It would mean coal pits the size of the Grand Canyon, the consumption of more coal than the US could possibly produce for several years, and the draining of crucial rivers like the Colorado to fuel the steam-dependent coal-shale plants and slurry pipelines. As one energy policy analyst put it, "Anything you can do with oil you can do with coal, only worse ... Overriding all the disadvantages (cost and technology) is the problem of air and water pollution." Carter's declaration included only five words on the question of the problems: "We will protect our environment." He did not say how.

As an EPA spokesman reluctantly admitted in reaction to Carter's declaration, "It may mean we won't solve the acid rain problem as quickly or as best we can. Yes, acid rain will increase." (The EPA had virtually no involvement in drafting Carter's plan and no advance warning of its announcement. The plan was formulated in conference with captains of industry, religious leaders and Democrats who had been summoned to Carter's mountaintop retreat at Camp David.) The declaration was Carter's most important statement to his nation since he had taken office, and—in those palmy days before Iranian crises, soaring inflation and industrial decline—was taken to be the cornerstone of his 1980 re-election platform. In Ottawa, Canadian environment officials voiced their dismay: "It's not going to make anything [an air pollution agreement] easier." Ontario's environment minister warned that the US energy plan could mean "new floods of acid rain for Ontario" if implimented without stringent controls. Even Canadian environment minister John Fraser conceded Carter needed to offer some "assurances" of environmental vigilance. But Fraser too was riding high on major commitments—Canada and the US were on the verge of signing a diplomatic agreement to set the stage for a full-scale treaty to combat acid rain; and he said he was confident that no US administration would sacrifice the environment. Indeed, he chided his officials for pessimism. Canada was moving fast on acid rain, Fraser insisted.

By early August the agreement to agree on the need for a treaty and promises to meet again had been agreeably signed; Carter had declared acid rain a serious global threat, in his annual environmental message—and promised $10 million per year over the next decade to study it; and John Fraser had returned from his first visit to Washington "warmly received by the very concerned" American interior and environment administrators Cecil Andrus and Doug Costle. They, like Canada, wanted an early achievement of a full-scale anti-acid rain treaty, Fraser reported. But behind the public warmth had been private sessions of stiff bargaining. Fraser had been forced to commit Ottawa to an effective Canadian abatement program even if it meant imposing federal regulations on reluctant provinces. "Canadians are producing a good part of the problem," Fraser conceded, and needed to bring their air pollution regulations more closely in line with the Americans'. But he remained confident: "It has been made clear that Carter has no intentions of abandoning environmental concerns," Fraser announced, in Washington before heading home.[3]

Such an assurance may have further bolstered Fraser's prominence in Canada as the tough environmental negotiator, and added strength to his pressure for a commitment to act against Ontario sources such as Inco, but in fact he had been misled, at best. As legislators, government officials, environmental groups and Canadian observers in Washington willingly admitted, the acid rain battle in the US was going badly. Largely forgotten in Carter's attack on energy, acid rain was unknown to the majority of legislators and citizens, and in fact many US administrators looked to Canada for a visible, inspiring initiative. These realities came clearer to Canada over the rest of 1979, but were ignored as first Ontario and then Ottawa shrank from any opening gambit in what was to remain a transboundary contest of you-go-first.

No one can better explain the realities of American interest in, and action against, acid rain than particularly informed Americans.[4] As a senior aide to then senior Democrat Senator Edmund Muskie (now Secretary of State) put it in mid-September, 1979 referring to the chance of an international treaty: "Acid rain is going to get worse. Don't expect any great initiatives in the near future. This is very much a time in America to circle the wagons and protect what legislation we have." The legislation included the weakened new source controls imposed in early 1979 after two years of podium promises and back-room deals. To do more, to impose controls on existing sources, would require new regulatory power for the Environmental Protection Agency. "Most utilities don't want to retrofit with scrubbers," the aide continued. "And most of the studies are biased against scrubbers. Yet we can't control acid rain without new legislation [to impose scrubbers]. But there isn't a great constituency for regulation these days. Bills are coming to Congress with awesome deregulation implication." One of those bills was part of Carter's go-coal policy package, which called for an Energy Mobilization Board empowered to bypass so-called red tape and time-consuming environmental assessments. "Right now in Congress all action is in response to outside pressures—pipelines, energy, and other interest-group demands." The aide was not hopeful. Perhaps after 1981 when the Clean Air Act automatically came up for Congressional review and possible improvement, sulphur dioxide reductions could be proposed for existing sources, "but I can't see anything happening in the next year or two." In fact, he, like other Capitol Hill veterans, warned of an equal possibility that any effort to toughen the laws could emasculate what there was, through the intricate Congressional system of trade-offs, lobbying and the national obsession with cheap energy.

A few blocks away in Washington at almost the same time, Representative Toby Moffett of Connecticut, chairman of the House subcommittee on Environment, Energy and Natural Resources, described the American situation this way:

> There is a body of opinion that is concerned with all serious sources of pollution, including those outside our border, and would like to see greater cooperation among nations, among Canada and the United States, to try and curtail things. At the same time there is also a powerful lobby within our own country to roll back the progress we've made on pollution and to see that we don't have any additional significant progress in terms of regulation. The lobby's tremendous roles in financing campaigns, their access to the public through advertising, hundreds of millions of dollars in advertising—it's pretty tough to go against that kind of machine.

Moffett was pessimistic. He didn't see "anything in the way of strong statements coming out of Congress. So many members regard it as hypocrisy—if we don't have the vote here to maintain our own laws, how can we go about urging other countries, even our neighbors, to do that? So I don't anticipate any [Congressional] dealings." As he had pointed out, earlier that day during committee testimony:

> Even though the Administration of the EPA and the President might have their hearts in the right place, the fact of the matter is that the environmental interests in this administration are no match for the [energy] interests and the Council of Economic Advisors and the Business Roundtable and the Political Action

Committees and the mythology that inflation is caused by government regulation .... Isn't it a question of the people at the EPA looking at this situation and saying 'we don't have the votes, we don't have the power, we don't have the leadership at the White House, we don't have the backing? ... It's a totally stacked deck. The monied interests are pushing in one direction and there is not enough pressure from the other side.

Karl Braithwaite, staff director of the Senate subcommittee on pollution, was equally negative about changing the prevailing pressures.

A lot of people are mildly aware of acid rain and that's good because a few years ago very few were. But in terms of aware to the point that it will stimulate some action, that hasn't happened yet .... A year ago we decided we ought to try on this subcommittee to have a hearing on acid rain some time within the year. But the Energy Mobilization Board has disrupted our schedule. We haven't had the luxury of some free time, but it is still one of the issues kicking around that we would like to have a hearing on in the next year or two, to get members familiar with it.

Across town, at the converted apartment building headquarters of the Environmental Protection Agency overlooking two abandoned slimy swimming pools, Doug Costle, administrator of the Environmental Protection Agency, was almost equally candid: "Control on existing sources (of sulphur dioxide) is the answer to acid rain. Talk about tall stacks is no answer." A tall, barrel-chested man who wears rumpled shirts and a loosened tie to meet reporters—and borrows their cigarettes to fuel his chainsmoking—Costle had earned a reputation as the most aggressive and effective environmental manager in the US since his appointment in 1977. Under Costle the EPA had begun a succession of detailed economic analyses to prove the financial benefits of cleaner air and water—and to defuse the corporate complaints of inflationary costs—and had salvaged much of the reputation of the largest regulatory agency in the country, while being demeaned and damned by industrialists and environmentalists alike. And yet, on acid rain, Costle said he was powerless. "The EPA cannot go out now and tell a typical 14-year-old coal-fired power plant in Ohio to install scrubbers. We just don't have the legislation. Congress told us to set ambient standards as the control for sulphur dioxide, and we did. But we just didn't know enough back in 1971 and 1977 to set standards for sulphates [acid rain]. Right now, it is up to individual states to set sulphate standards," he added, without mentioning that no state had done so.

"The Canadian complaint on acid rain is a valid one—we're still sending 4 million tons of sulphur dioxide up there. And if their information about 48,000 lakes dying is valid—we're checking their data—they've got a right to scream. I think what's going to happen is that acid rain will dominate Canada-US relations, and the US Clean Air Act Amendments of 1981," he said, but he would give no indication of what amendments he wanted, or what outcome he anticipated. "It may mean we have to put scrubbers on existing plants, despite the costs," was the limit of his analysis.

Costle's staff offered more details. "Acid rain will continue to fall, to at least the end of this century, over a wider and wider area," said Dennis Tirpak, the EPA's chief of acid rain research in a subsequent interview. Tirpak cited evidence of crop damage—"a critical situation"— and the futility of liming lakes and the impossibility of liming forests. He mentioned health damage from acid rain. He conceded that the US had been slow to recognize the problem and still lacked good measurements; he admitted the lack of any regulations governing acid rain sulphates; and agreed that it would take some reasonable estimates and calculations of the cost of acid rain to Americans to wake them up to the need for more pollution controls. "It's going to take three to five years to get a good cost-benefit analysis" to counter industrial claims of overburdening costs of abatement, Tirpak said, lamenting the far too limited previous funding for acid rain research. But Tirpak, aided by Konrad Kleveno, the EPA's chief of Canada-US relations, could not miss the opportunity to refer to Canada too. "We're concerned that Canada has no controls on its sources going into operation—Atikokan, Poplar River, and the Ontario Hydro Nanticoke plant whose tall stack blows pollution directly to the Buffalo area we've worked hard to clean up." Both officials admitted that "the Carter energy plan is going to boost sulphur dioxide," but added that "it is also a matter of Canada getting its own house in order."

Barbara Blum, Costle's deputy and a very political lady, was even more direct. She, like Costle, argued in favor of tough controls on existing sources but emphasized the EPA's inability to impose them. "We're just getting a handle on how serious the problem is. Yes it is true we can't afford to wait, we may have to go back to Congress for more power, but I'm not sure we could force on the states a program to force controls on existing sources. It is a very difficult bullet to bite." Blum however insisted "Carter's policy has never changed. We'll have more coal energy and we'll do it safely." (Blum had spent hours only a few days previously assuring a Congressional committee that environmental regulations would not delay the growth of US energy supplies.) And she finally added, "You Canadians have got a real problem up there, with the biggest source in the world [Inco]. It looks like your federal government has to get some power to press the provinces to act."

On September 11, Robert Rauch, staff attorney for the Environmental Defense Fund, one of the largest, most powerful and most competent US environmental organizations, had preceded Ms. Blum in testifying before the energy, environment and natural resources subcommittee of the House of Representatives, chaired by Toby Moffett of Connecticut. Rauch's testimony garnered careful attention by the committee, perhaps partly because he was a former assistant to the deputy administrator of the EPA. Rauch described the futility of imposing only new source controls when old sources remained unhampered—90 per cent of sulphur dioxide power plant emissions would still come from plants built prior to 1970, with no controls at all, in 1985; he noted that a typical Gavin, Ohio, power plant built without controls could legally emit eight times more sulphur dioxide than a new, controlled one built beside it—hardly an incentive for a company to phase out old plants early. Rauch cited an academic study which concluded that the combined annual damages of a single 500 megawatt uncontrolled coal plant might range from $7 to $50 million, including lake, property, vegetation, visibility and human health damage. But "despite these statistics, the pressure on the EPA to further relax emission limitations for existing power plants continues to grow," Rauch warned. He cited Energy Department requests to the EPA to reconsider whether the limits imposed on power plants were not too tough. Some revisions had already been made which resulted in boosted emissions. Other individual sources, particularly plants in Ohio, the single worst air polluter in the US, were pressing for further relaxations.

Rauch described how the Cleveland Electric Illuminating Company had sought a 400 per cent increase in emissions from its two plants at Avon Lake and Eastlake. Local United Mine Workers were demanding that local Ohio coal be burned by Cleveland Electric—coal so dirty it would absolutely require scrubbers installed on the two plants. The company claimed the existing limits were inaccurately calculated and that relaxation—by 400 per cent—was needed, and thus no scrubbers. Somewhere between April and mid May, 1979, following top-level EPA meetings, the agency officials in Ohio opposing the relaxation application suddenly reversed their stance and granted it, without explanation. The decision took the EPA and the White House—both of whom denied influencing the local decision—"off the hot seat," said Rauch. "The agency would presumably have had to decide between requiring Cleveland Electric to put on scrubbers in order to continue burning high sulphur Ohio coal or risk the wrath of the coal industry by allowing the utility to go out of state and purchase low sulphur coal." It was only one example, more applications were pending; there was also the looming threat of a resumption of tall stacks. Nor was the EPA as powerless as some claimed in combating long-range transport of acid rain precursors. There was a section of the Clean Air Act which could ban emissions from one state which fouled any downwind state's own efforts at cleaning up its air. "Up to now, the EPA has steadfastly ignored these requirements," Rauch commented, although Cleveland Electric's increased emissions were already under attack from downwind New York and Pennsylvania. "EPA has the statutory tools which are needed. What it lacks is the political will to use them." In the US the acid rain battle was going very badly indeed.

None of this was unknown to at least some Canadians, including those who accompanied and greeted John Fraser when he first came to Washington one month earlier and proclaimed he was assured of no relaxation in environmental protection for the sake of energy. As a senior official at the Canadian embassy in Washington assessed the situation: "The Administration, the executive, the White House, the agencies like the EPA, and the people in the Congress—if they manage to hold the line on major environmental issues and give ground only in a very limited way, on some of the not-crucial issues, then they will have done very well .... It is a time of passing laws to make exceptions of the general laws .... In the short run there are two overriding circumstances. One is the presidential election and the other is what a lot of people perceive to be an energy crisis. And those two things together have a lot to do with the fact that the EPA have their heads down .... There is nothing even vaguely approaching consensus on how to deal with the energy crisis." And on acid rain, the increased product of that go-coal energy declaration of Jimmy Carter: "There is not a great awareness of it."

But from those who were aware, the rare politicians, the administrators and environmentalists, the Canadian official had gained a critical perspective:

> Basically what they are saying—and this is among people who are favorable to a treaty—is "if you guys [Canadians] think you are going to have a snowball's chance in hell of moving this treaty thing forward, given the political climate, then we must have at the very minimum proof, not just promises, that Can-

ada and in particular Ontario, is prepared to take tough measures in areas where it appears that south of the border they're taking the easy road." Those examples include the Sudbury smelter, with the publicity that has come along about the environment ministry saying one thing and backing off. And they have obviously mentioned Atikokan.

This too was known by Ottawa, and Ontario, where the environment ministers were already reaffirming their policies of taking no precipitous abatement action against acid rain sources before the ostensibly aware and concerned Americans did likewise. Both sides were settling in for a long rain. At least some Americans were prepared to admit it.

A few days later, in Ottawa, Canadian environment minister John Fraser offered what he called his "most candid" assessment of the acid rain problem, in an interview.[5] Acid rain was the highest priority for his department, backed by Prime Minister Joe Clark. "The result of doing nothing is totally unacceptable for Canada. We must act. I concede that there have been allusions and cynical comments about the difficulties of getting the US to act, what with their own problems and the ramifications of the upcoming presidential election and an administration going on HOLD. But I seriously don't believe any American political party would seriously consider dumping environmental concerns." Fraser talked of acid rain as an issue of such seriousness it could not be shelved—it would affect all Canada-US relations. "I can't see the Americans ignoring us, we have too long a history of seeking common resolution to common problems. ... The Americans need us, our ideas, our resources, our people. There will be an explosion of Canadian anger the likes of which our neighbor has never seen if we have to wait 15 years for abatement and thousands of lakes to die. The Americans cannot afford to have a hostile Canada."

It was tough talk, although Fraser refused to discuss what direct pressures Canada could wield, and conceded that on auto pacts and foreign ownership Canada appeared weak. Nor would he put a date on the accomplishment of an effective treaty. He said initial abatement action by Canada was one option worth considering: "I'm counting on the co-operation of Ontario and Quebec, as opposed to direct federal action. Ontario environment minister Harry Parrott has given me his word that Ontario will not be a stumbling block. But I'm not pressing Ontario to clamp down on Inco, not this month."

But finally, Fraser added, "I admit we must hold out hope. We can't just let the issue resolve itself in time. This is my challenge, to keep Canadians aware they are facing a decade or more of acid rain if we don't press for action." Two weeks later at an air pollution conference in Ottawa, Fraser announced he would not impose a Canadian clean-up of acid rain sources until it was matched by simultaneous American effort. Canadian industry could not afford to be pushed to the same level of emission controls, particularly on new sources. "At this time my strategy is to move equally with the Americans," he said. Simultaneously in Washington, the EPA put out its acid rain assessment: it was going to get "significantly worse" to the end of the century.

In mid October, the first joint Canada-US acid rain study was released, with predictions of widespread lake damage and ominous warnings of crop, forest and property losses. It held Canada responsible for about half of the total acid material deposition it suffered. The subject mushroomed in media attention and was even enshrined in a rock song by a Toronto group (Joe Hall and the Continental Drift). Demands for specific commitments to action by both the federal and provincial governments mounted, and Ontario environment minister Harry Parrott opened his routine of on-again-off-again promises to impose a new control order on Inco. His premier made it clear any such plans were off. At the November 2-3 international Action Seminar on Acid Precipitation in Toronto Parrott re-emphasized Ontario's intention to "work in step with other jurisdictions, hopefully through an international agreement," although he insisted Ontario would take action at some unspecified time on an individual basis too. John Fraser said he feared no international treaty was possible until after the post 1980 US presidential elections, but Canada would take action, treaty or not, within two years.

Congressman James Oberstarr, of the acid-plagued state of Minnesota and one of only two American political representatives at the conference, had the last, and possibly the most prophetic words: "The mood of the US Congress is not good. Acid rain is more pervasive and more destructive than any other phenomenon we've released this century. There's never been a time when we could look ahead more clearly and see a disaster coming." By November, 1979, the long hot summer of acid rain had turned to winter, and soon the rain would turn to acid snow, to accumulate as acid shock loadings for Canadian rivers and streams in 1980. And Ottawa and Washington announced their agreement to hold another meeting, in the near future, to try to agree on an accelerated timetable for action.

# 10
# Grim Forecast

Harold Harvey published his discovery of acid death in the Killarney Lakes in 1971. Sweden released its "red book" in 1971 which termed the acid-rain situation "disastrous" in Scandinavia and likely to be duplicated in Canada and the northeastern United States. Harvey was almost alone in his apprehension; the Swedish reports were virtually ignored. It's taken nearly a decade of cautious, uncommitted, uncoordinated research to confirm that Harvey and the Swedes were unerringly accurate. Today the scientific reports are no longer so discreetly speculative but grow increasingly alarmist. Acid rain is no longer a subject for technical journals and academic conferences only. Now newspapers and magazines, including those where Canada still advertises her pristine north, feature cartoons of citizens holding disintegreated umbrellas in the rain or catching skeletal fish. The essential physical facts of acid rain are being established in the minds of everyone.

The financial realities are also gaining publicity. The cost of reducing the culprit emissions will be expensive, but the corporate and political claim—some call it blackmail—that the costs are unacceptably high and threaten industry collapse may soon lose its intimidating power. The technologies for abatement are available, and given the size of most of the corporate polluters, affordable. More importantly, as the belated estimates of the cost of acid rain damage are tallied, it will become evident that we simply can't afford it. In fact, the cost escalates with every day's delay in applying the abatement technologies, and scientists now warn the damage is nearing an irreversible stage with most dire economic consequences.

What remains unchanged is any political commitment to face these realities and begin turning off the acid rain now. What remains unrecognized too is that the cost of acid rain is being involuntarily borne by every North American. Those living in the northeast, particularly Canada, are only first in line to pay the inevitable consequences of a corroded environment, destroyed fishing, and deflated recreation and tourism industries. Forestry, agriculture, property and possibly human health costs will be spread far wider.

The political response to acid rain in the 1970s was dominated by a shortsightedness and deceptive posturing which has remained largely unchanged throughout the decade. Judging by the policies and tactics exercised in a series of important events early in 1980, the long-range forecast remains grim.

● In mid January, 1980, Canada and the US announced a joint $100 million commitment to study acid rain. Canadian environment minister John Fraser said he remained optimistic that an effective abatement treaty could be signed within a year. Three weeks later it was revealed that Jimmy Carter wanted to make more acid rain.[1] He wanted at least 62 oil-fired power plants to convert to coal, quickly. Largely exempt from the tight pollution controls required on all new plants, the conversions would reduce reliance on oil, and increase sulphur dioxide emissions by at least 600,000 tons per year. Most of the plants are located in the already acid-saturated northeast. The conversion plan had already been outlined to utility and Congressional leaders, to garner their support. The Environmental Protection Agency had been among the last to be consulted. Canada had been told nothing.

● In Ottawa in mid February John Fraser said he was "concerned" about the US conversion plan. He said he was pushing the Americans to sign a treaty which would include tight controls on the converted plants too, and would demonstrate American good faith in the fight against acid rain.[2] One week later Fraser was pushed out of office, his Conservative party defeated after only six months in power, the victim of its intention to govern from a minority position as if it held majority power. In Washington the Carter conversion scheme went to Congress without tight controls on emissions. The Carter administration recognized acid rain, a spokesman said, but acid rain was "a new phenomenon which has not been resolved." It could not stand in the way of rapid conversion to more coal use.[3]

• John Roberts entered office, the fourth Canadian environment minister in as many years, a Liberal again, yet like his Conservative predecessor, John Fraser, an attractive, ambitious man, determined to raise his profile by attacking acid rain. He declared it an "environmental time-bomb," asserted that Canada was prepared to reduce its own emissions drastically, and insisted that the provincial governments were in agreement.[4] Meanwhile, Noranda Mines Ltd. warned that it would probably have to close down its giant Rouyn-Noranda smelter if the Quebec government insisted on substantial reductions of sulphur dioxide emissions.

• In early April Ontario environment minister Harry Parrott suddenly flip-flopped again, and announced he would "do something independently of the US" to begin a crackdown on Ontario emitters. Two weeks later he declared he was opposed to Ontario Hydro's plan to burn more coal for the sake of producing more power to sell to the US. (Ironically, Hydro wanted to sell the power to General Public Utilities Corp. whose Three Mile Island nuclear power plant was permanently out of order.) Hydro's export scheme would also produce another 80,000 tons per year of sulphur dioxide from its unscrubbed smoke stacks. But Ontario premier William Davis quickly corrected Parrott. It was preferable for Hydro to burn more coal and increase its revenue by exporting the power—and the emissions—south to the US, than to have the Americans do it for and to themselves, declared the premier, long known for his boosterism of Hydro and not known for rabid environmentalism or understanding of acid rain.[5]

• In late April John Roberts said the Americans had

convinced him. Something had to be done about Canadian polluters first, and Inco was the prime target. Inco, he said, could reduce its Sudbury emissions 50 per cent, using existing technology, within five years, at a cost of about $400 million. He said he had new federal studies to prove it. It would add 20 cents per pound to Inco's production costs for nickel (then $3.50 per pound), a "not profitable but affordable cost for Inco." The giant multi-national company was the most financially healthy nickel producer in the world, with the lowest production costs, and capable of absorbing $400 million expenditures without government aid.[6] Harry Parrott met with Roberts, and said afterward he was in "general agreement." A new control order for Inco was in preparation. Inco replied that "it is misleading to suggest that reducing Inco's emissions will have any significant effect on the problem of acid rain."[7]

● On May 1, 1980 the new control order to "substantially reduce" Inco's sulphur dioxide emissions was announced. It would "set an example and lead in responding to acid rain," promised Parrott.[8] In fact, the order gave Inco two more years of uninterrupted pollution, imposed a 25 per cent reduction by 1983, and requested an unspecified further reduction at an unspecified later date. The order immediately froze Inco emissions at 2,500 tons per day; but Inco had been emitting almost exactly that amount for many months because world nickel demand was down, production was down (although nickel prices and Inco profits were up.) The order required a 25 per cent reduction in two years; but Inco had announced two months earlier that it had designed a new process which would reduce emissions 25 per cent, and if perfected could be installed in two years. It would cost some "tens of millions of dollars," but Inco's profits had risen from $78 million in 1978 to $141 million in 1979.

The substantial reduction promised by Parrott remained undefined and unscheduled. The only target referred to was "as low as possible" which could mean anything depending on Inco's definition of what was possible and neither unprofitable nor unfeasible. Inco and a federal-provincial task force of civil servants were to study the question, together. No target date was set, no reference was made to the federal studies which talked of a 50 per cent reduction within five years at a cost of $400 million. In the Ontario Legislature Opposition critics called the order a "sell-out" and "written by Inco." In Ottawa John Roberts praised Parrott for his "important and courageous decision." In Toronto Inco chairman Charles Baird conceded the control order posed no immediate inconvenience for the company, but if nickel demand suddenly skyrocketed the company might be frozen out of as much as $700 million extra income per year (based on boosting production of nickel, and matching emissions, back to the 3,600 ton per day level).[7] He also warned that if the promised 25 per cent reduction technology failed, after 1982, Inco might be forced to cut pollution 25 per cent by cutting nickel production, and that, he reminded reporters, could mean cutting jobs in Sudbury.

• In early May Ontario environment minister Harry Parrott promised an announcement of abatement strategies for Ontario Hydro's emissions within a few weeks. A few weeks passed and nothing was announced. The Ontario Legislature adjourned for the summer. In Ottawa John Roberts' parliamentary secretary Roger Simmons said Ontario's action against Inco was only part of the solution to acid rain. He did not suggest any of the parts included the Noranda smelter in Quebec, the other Inco smelter in Manitoba, the projected coal-fired power plants on the prairies, the predicted 30 per cent increase above 107,000 tons per year in Alberta as new oil sands and refining plants were constructed, or British Columbia's intention to use uncontrolled coal-fired plants instead of nuclear power. He said the next part was convincing the US government to impose tight controls on Jimmy Carter's oil-to-coal conversion plan. "The light at the end of the tunnel is that Mr. Carter's coal conversion program has not passed the Congress," Simmons said.[10] He ignored the more than 200 existing US coal-fired plants under no controls and 350 new plants which presaged an increase in total emissions. The tunnel

Simmons referred to was in fact a giant funnel bearing more acid rain to Canada.

    • On June 25 the Carter conversion scheme passed through the US Senate without imposition of tight sulphur dioxide emission controls. US Environmental Protection Agency chief Doug Costle had long given up the fight in that small skirmish and was looking ahead. Acid rain was going to worsen, he told a Senate subcommittee.[11] Its impact was already recognizable from Minnesota to Pennsylvania to the Great Smoky Mountains (an increasingly appropriate name) in acidic waters, fish mortality, toxic drinking water, and initial confirmations of crop damage. Costle was optimistic about techniques to reduce acid rain—scrubbers, coal-washing, low-sulphur coal—but he was pessimistic about the application of those techniques to the acid rain sources. It could take up to 10 years to develop and impose the regulations needed to deal with acid rain, he said. Meanwhile, the state of Pennsylvania launched a lawsuit against the EPA for relaxing regulations allowing Ohio to boost its emissions, and New York and New Jersey went to court to force the EPA to relax *their* regulations.[12] As Canadian officials confirmed, nobody in Washington noted Canada's lone initiative against Inco as the break-through in the transboundary deadlock.

    • But to hear the coal lobby tell it, there is no cause for concern. "It is too early and not substantiated that the burning of coal is a source of acid rain," declared Jack Keany of the utility industry's Edison Electrical Institute.[13] "Considerable national anxiety about acid rain has been stirred up by the EPA and the national media," commented Carl Bagge, president of the National Coal Association. "But solid information is not available to justify the imposition of regulations or controls."[14] The American Electrical Power System which operated coal-fired plants in seven eastern states announced it would conduct its own study "to determine the causes and effects, whether beneficial or detrimental, of acid rain."[15] As William Poundstone, vice-president of Consolidation Coal Company instructed the Senate energy and natural resources committee on

May 28, "It is not clear that acid precipitation does in fact cause acidification of lakes ... It is also unclear that coal burning by utilities is a major cause of increased acid rain ... It would be unwarranted, unjustified and unwise for the nation to embark on a course of regulatory controls based on scant, conflicting and inconclusive data."

The litany of promises, veiled threats and futile gestures goes on. In June, 1980, Inco officials declared it very doubtful that the company could accomplish its 25 percent reduction in pollution emissions at the Sudbury Superstack within two years as previously promised by the company and threatened by the Ontario government. The company said it had too little time to work out its technical problems.[16] By August the government had failed to impose the control order and was gearing up for a lengthy period of appeals by Inco against the order, whenever finally imposed. Canada and the US signed a "memorandum of intent" to fight acid rain together by trading information, enforcing existing laws, and of course studying the problem further. The pledges differed little from ones announced a year earlier, aside from an agreement to begin formal discussions about a transboundary air treaty—in another year. Meanwhile, US legislation to hastily convert oil-fired power plants to coal, exempted from pollution controls, neared approval by Congress. And in Sweden and the US, experimenters announced enthusiastic plans to breed an acid-resistant fish, capable of withstanding low pH and heavy aluminum dosages.

Harold Harvey, the Canadian zoologist who first found acid—and no fish—in the Killarney Lakes of Ontario a decade earlier, has an appropriate assessment of the resistant fish scheme. Fish, he points out, are only one measure of the impact of acid rain, serving as an early warning of impending ecosystem death much like canaries in cages taken into mines as a warning against poisonous gasses. Breeding acid resistant fish is thus no wiser than equipping canaries with gas masks.

Acid rain is an environmental crisis with sweeping financial implications, but its solution is political. In the US the politicians are making laws and finding ways around them; in Canada they still seek to avoid making laws concerning acid rain. If the political tradition, the claims of the corporations, and the quiescence of the scientists to date represent the latest words on acid rain, then the long range forecast is an epitaph for the unprotected environment of North America. It doesn't have to be that way. The future remains the choice of every North American.

*—Toronto, August 1980.*

# End Notes

## 1 The Dying Lakes

1. Beamish, R.J. and H.H. Harvey, "Acidification of the La Cloche Mountain Lakes, Ontario, and Resulting Fish Mortalities," *Journal of the Fisheries Research Board of Canada*, Vol. 29, 1972, 1131-1143.
2. Ontario Ministry of the Environment, *Effects of Acid Precipitation on Precambrian Freshwaters in Southern Ontario*, April, 1978.
3. Ontario Ministry of the Environment, *Legacy*, Vol. 8, No. 1, 1979.
4. Ontario Ministry of the Environment, *Extensive Monitoring of Lakes in the Greater Sudbury Area, 1974-1976*, 1977.
5. Hendrey, George R., *Acidification of Aquatic Ecosystems: Ecosystem Sensitivity and Biological Consequences*. A presentation to the Action Seminar on Acid Precipitation (ASAP), Toronto November 2, 1979.
6. Grahn, O. et. al., "Oligotrophication—A Self-Accelerating Process in Lakes Subject to Excess Acid," *Ambio*, Vol. 3, No. 2, 1974.
7. Brodde, Almer et al. in *Sulphur in the Environment: Part II: Ecological Impacts*, J.O. Nriagu, ed., John Wiley & Sons Inc., New York, 1978; and Kronen, C.S. and C.L. Schofield, "Aluminium Leaching Response to Acid Precipitation: Effects on High-Elevation Watersheds in the Northeast," *Science* 204, 1979: 304-305.
8. Schofield, C.L., "Forecast: Poisonous Rain," *Saturday Review*, February 9, 1978.
9. Brouzes, R.J. et al., *The Link Between PH and Mercury Content of Fish*. A paper presented at the U.S. National Academy of Sciences National Research Council Panel on Mercury, Washington, D.C., May 3, 1977.
10. Likens, Gene, "Acid Precipitation," *Chemical and Engineering News*, November 22, 1976.

## 2 Acid in The Air

1. See Likens, Gene, "Acid Rain," *Scientific American*, October, 1979; and Cogbill, C.V., "The History and Character of Acid Precipitation in Eastern North America," *Water, Air and Soil Pollution 6*, 1979; and Granat, Lennart, "On the Relation Between PH and Chemical Composition in Precipitation," *Tellus XXIV*, 1972, 6.
2. Shenfeld, L. et al., *Long-Range Transport of Ozone into Southern Ontario*, Ontario Ministry of the Environment, 1979.
3. Summers, P.W. and D.M. Whelpdale, "Acid Precipitation in Canada," *Water, Air and Soil Pollution 6*, 1976.
4. Nordo, J., "Transport of Air Pollutants in Europe and Acid Precipitation in Norway," *Water, Air and Soil Pollution 6*, 1976.
5. Glass, Norman, *Acid Precipitation in the U.S.: History, Extent, Sources, Prognoses*, Interim Report, Corvallis Research Laboratory, U.S. Environmental Protection Agency, 1979.
6. Millan, Millan and Y.S. Chung, "Detection of a Plume 400 kms. from its Source," *Atmospheric Environment*, Vol. II, 1977.
7. McBean, G.A., Opening remarks to ASAP, Toronto, November 2, 1979.
8. U.S.-Canada Research Consultation Group, *The LRTAP Problem in North America: A Preliminary Overview*, October, 1979.

## 3 Acid in the Eco-system

1. Whittaker, R.H. et al., "The Hubbard Brook Ecosystem Study: Forest Biomass and Production," *Ecological Monographs*, Vol. 44, No. 2, 1974.
2. Hutchinson, T.C. & L.J. Whitby, "The Effects of Acid Rainfall and Heavy Metal Particulates on a Boreal Forest Ecosystem Near the Sudbury Smelting Region of Canada," *Water, Air and Soil Pollution 7*, 1977: 421-438.
3. Jonsson, Bengt, "Soil Acidification by Atmospheric Pollution and Forest Growth," *Water, Air and Soil Pollution 7*, 1977: 500.
4. Dochinger, L.S. and T.A. Seliga, Introduction to "Acid Precipi-

tation and the Forest Ecosystem: Report from the First International Symposium," *Journal of the Air Pollution Control Association*, Vol. 25, No. 11, 1975: 1103-1105.

5. Cowling, E.B. "Effects of Acid Precipitation on Terrestrial Vegetation," and other authors in *A National Program for Assessing the Problem of Atmospheric Deposition (Acid Rain)—A Report to the Council on Environmental Quality*, National Atmospheric Deposition Program, Natural Resource Ecology Laboratory, Colorado State University, 1978.

6. Denison, R. et al., "The Effects of Acid Rain on Nitrogen Fixation in Western Washington Coniferous Forests," *Water, Air and Soil Pollution 8*, 1977: 21-34.

7. U.S.-Canada Research Consultation Group, op. cit.

8. Tamm, C.O., "Acid Precipitation: Biological Effects in Soil and on Forest Vegetation," *Ambio*, Vol. 5, No. 5-6, 1976.

9. National Atmospheric Deposition Program, op. cit.: 64-73.

10. Hindawi, I.J. et al., "Response of Bush Bean Exposed to Acid Mist," *American Journal of Botany 67*, 1980: 168-172.

11. Pough, E.H., "Acid Precipitation and Embryonic Mortality of Spotted Salamanders," *Science 192*, 1976: and Brodde, op. cit.

## 4 Acid in The Community

1. Oden, Svante, "The Acidity Problem: An Outline of Concepts," *Water, Air and Soil Pollution 6*, 1976: 137-166.

2. Swedish Royal Ministry for Foreign Affairs and Royal Ministry of Agriculture, *Air Pollution Across National Boundaries—The Impact on the Environment of Sulfur in Air and Precipitation—Sweden's Case Study for the United Nations Conference on the Human Environment*, August, 1971.

3. Fuhs, G.W., *A Contribution to the Assessment of Health Effects of Acid Precipitation*, N.Y. State Department of Health, Albany, N.Y., 1979.

4. Legislative Assembly of Ontario, *Hansard*, Vol. 3, No. 69, June 12, 1979: 2800.

5. Mendelsohn, R. and G. Orcutt, "An Empirical Analysis of Air Pollution Dose-Response Curves," *Journal of Environmental Economics and Management*, Vol. 6, No. 2, 1979: 85-106.

## 5  Acid Economics

1. Ruston/Shanahan & Assoc. Ltd., Hough, Stansbury & Assoc. Ltd., and Jack B. Ellis and Assoc. Ltd., *The Fishing and Hunting Lodge Industry in Ontario*, Report prepared for the Ontario Ministry of Northern Affairs and the Northern Ontario Tourist Outfitters Association, January, 1979.
2. Liddle, Jerry, *Potential Socio-Economic Impacts of Acid Rain*. A presentation to ASAP, Toronto, November 2, 1979.
3. See Ontario Ministry of Industry and Tourism, *The Economic Impact of Tourism in Ontario and Regions, 1976*, and *Framework for Opportunity: A Guide for Tourism Development in Ontario/ Canada, 1977*, and *Ontario Ministry of Industry and Tourism Review, 1978-79*.
4. Swedish Royal Ministry of Agriculture, *Draft Estimation of Financial Damage to Aquatic Systems*, Room Document No. 16, October 16, 1978.
5. Roberts, Hon. John, Speech to the Canadian Pulp and Paper Association, Quebec City, March 25, 1980.
6. See Nriagu, J.O., op. cit., Vol. II, for more than 200 references to property damage.
7. See Ruby, M.G., *An Application of Cost-Benefit Analysis: The Arasco-Tacoma Copper Smelter*. A presentation to the Annual Meeting of the Pacific Northwest-International Section of the Air Pollution Control Association, Portland, Oregon, November, 1978.

## 6  Weak Laws and Big Bucks

1. Wetstone, Gregory S., *Air Pollution Control Laws in North America and the Problem of Acid Rain and Snow; Environmental Law Reporter*, Washington, D.C., February 1980, No. 10: 50001-50020.
2. Cited by Gus Speth, Chairman, President's Council on Environmental Quality, at the Action Seminar on Acid Precipitation, Toronto, November 2, 1979.
3. U.S. Senate Committee on Energy and Natural Resources, *Interagency Task Force Report: Sulphur Oxides Control Technology in Japan*, Washington, D.C., June 30, 1978.

4. Costle, Douglas, "New Standards for Coal-Fired Power Plants," U.S. Environmental Protection Agency (EPA) Media Release, Washington, D.C., May 25, 1979.
5. *Steam-Electric Plant Air and Water Quality Control Data for the Year Ended December 31, 1975*, Federal Energy Regulatory Commission, U.S. Department of Energy, Report No. DOE/ FERC—0024.
6. Personal communication from William Baasel, U.S. EPA, Research Triangle Park, North Carolina, May 16, 1980.
7. Gouvernement du Québec, Services de protection de l'environnement, Bureau d'étude sur les substances toxiques, Projet "Région Rouyn-Noranda," *Group "Intervention": Document de Consultation*, Rapport I-7, juillet, 1979.
8. Lemmon, Bill, Air Pollution Control Directorate, Environment Canada, in an interview with Ross Howard, May 5, 1980.

## 7  Acid Politics in Ontario

1. Ontario Ministry of the Environment, *Impacts of Air Pollutants on Wilderness Areas of Northern Minnesota: Review of Final Draft Copy*, May 10, 1979.
2. Government of Ontario, *The Report of the Royal Commission on Electric Power Planning*, February, 1980.
3. Ross, Val, "The Arrogance of Inco," *Canadian Business*, May, 1979.
4. Statement by Inco Ltd. to the Standing Committee on Resource Development Legislative Assembly of Ontario, February 6, 1979.
5. Toronto Star, February 5, 1979.
6. Toronto Star, August 5, 1978.
7. "Plan to Dry up Acid Rain: Sell Smog," Toronto Star, February 16, 1980.
8. Mineral Policy Sector, Energy, Mines and Resources Canada, *A National Sulphur Strategy*, Report prepared by G.H.K. Pearse, Third Draft, October 24, 1979.
9. Energy Mines and Resources Canada, *A Marketing Study for By-product Sulphuric Acid*, Report prepared by the British Sulphur Corporation, 1979.

10. Air Pollution Control Directorate, Environment Canada, *Preliminary Assessment of Feasible SO₂ Emission Reductions*, Ottawa, 1980.
11. Globe and Mail, February 9, 1980.

## 8   The Ottawa Connection

1. Ontario Ministry of the Environment, July 31, 1978.
2. Toronto Star, July 21, 1979.
3. Toronto Star, July 17, 1979.
4. Globe and Mail, August 18, 1979.
5. Toronto Star, October 16, 1979.
6. Globe and Mail, October 17, 1979.
7. Globe and Mail, October 22, 1979.
8. Globe and Mail, October 26 , 1979.
9. Globe and Mail, October 27, 1979.
10. Toronto Star, June 21, 1977.
11. Parliament of Canada, *Hansard*, February 8, 1979: 3016.
12. Toronto Star, March 10, 1979, and personal interviews.
13. *Ibid.*
14. La Presse, June 2, 1979.
15. Halifax Chronicle-Herald, July 12, 1979.
16. Ottawa Citizen, July 13, 1979.
17. Toronto Star, July 21, 1979.
18. Globe and Mail, August 3, 1979, and CTV National Television News, August 9, 1979.
19. CBC Radio, "The World at Six," August 9, 1979.
20. Toronto Star, October 16, 1979, and personal interview, September 21, 1979.
21. Globe and Mail, September 26, 1979.
22. Globe and Mail, October 17, 1979.
23. Toronto Star, November 12, 1979.

## 9 South of the Border

1. Washington Post, May 4, 1979.
2. See "Can We Afford U.S. Energy Plan," Toronto Star, July 21, 1979.
3. Washington Post, August 13, 1979.
4. Toronto Star, October 16, 1979, and personal interviews.
5. *Ibid.*

## 10 Grim Forecast

1. Washington Post, February 7, 1980.
2. CBC Radio, "Daybreak," February 14, 1980.
3. Sommerville, Glenn, Canadian Press, March 3, 1980.
4. Speech by Ray Robinson, Assistant Deputy Minister, Environment Canada, to EPA conference on acid rain, Springfield, Virginia, April 8-9, 1980.
5. Globe and Mail, April 26, 1980.
6. Toronto Star, April 22, 1980.
7. *Ibid.*
8. Ontario Ministry of the Environment, News Release, May 1, 1980.
9. Inco Ltd., Media Information, IN 46/80, May 1, 1980.
10. Parliament of Canada, *Hansard*, May 6, 1980: 788.
11. Globe and Mail, March 20, 1980.
12. Bender, Judith, *Newsday*, April 16, 1980.
13. CBC National Television News, April 17, 1980.
14. *Locomotive Engineer*, Washington, D.C., May 2, 1980.
15. Globe and Mail, May 10, 1980.
16. Toronto Star, June 5, 1980.

# Resource List

## AGENCIES AND PEOPLE

### United States

The government source in the U.S. for acid rain information is of course the Environmental Protection Agency. Go through their office of Public Awareness, 401 M Street S.W., Washington, D.C. 20460, telephone (202) 755-2673 or 755-0344. The overall Agency Administrator is Douglas Costle.

Another major government agency is the Council on Environmental Quality, 722 Jackson Place N.W., Washington, D.C. 20006, telephone (202) 395-5700. The Chairman is Gus Speth, who has made some strong statements on the need for abatement.

Public interest groups deeply involved in the acid rain debate in the States include the Center for Law and Social Policy, 1751 N Street N.W., Washington, D.C. 20036, telephone (202) 872-0670 (Jim Barnes);

The Environmental Defence Fund, 1525 Eighteenth Street N.W. Washington, D.C. 20036, telephone (202) 833-1484 (Bob Rauch);

And the Environmental Law Institute, 1346 Connecticut Ave. N.W., Washington, D.C. 20036, telephone (202) 452-9600 (Greg Wetstone).

There are many more U.S. groups involved, but the above can put you in touch with them.

U.S. companies to contact regarding their SOx/NOx emissions and abatement efforts include American Electric Power Company, Commonwealth Edison, The Tennessee Valley Authority, and Detroit Edison.

### Canada

In Canada, a major participant at the federal civil service level in acid rain abatement is Ray Robinson, Assistant Deputy Minister, Environmental Protection Service, Environment Canada, Place Vincent Massey, Ottawa, Ontario K1A 1C8, telephone (613) 997-1575.

The present (July 1980) federal Minister of the Environment is the Hon. John Roberts, Minister of the Environment, 14th Floor, Fontaine Building, Ottawa, Ontario K1A 0H3, telephone (613) 997-1441.

The current (July 1980) Ontario Minister of the Environment is Dr. Harry Parrott, Minister of the Environment, 14th Floor, 135 St. Clair Ave. West, Toronto, Ontario M4V 1P5, telephone (416) 965-1611.

Senior Ontario civil servants to contact would be (a) Ed Piche, Coordinator, Acid Precipitation Study, Ministry of the Environment, 6th Floor, 40 St. Clair Ave. West, Toronto, Ontario, telephone (416) 965-1140;

(b) Dr. Greg Van Volkenburgh, Air Resources Branch, Ministry of the Environment, 4th Floor, 880 Bay Street, Toronto, Ontario, telephone (416) 965-2053 and

(c) Dr. Tom Brydges, Supervisor, Limnology and Toxicity Section. Ministry of the Environment, Resources Road, Rexdale, Ontario, telephone (416) 248-3058.

The former two are involved in the overall Ontario abatement effort, while the latter is particularly concerned with acid rain's impact on lakes and waterways.

As concerns public interest groups, the major one is the group which followed up on the November 1979 conference on acid rain held in Toronto. Contact the Federation of Ontario Naturalists, 355 Lesmill Road, Don Mills, Ontario M3P 2W8, telephone (416) 444-8419. Talk to Ron Reid for details on this group. The FON distributes a 27 page bibliography especially prepared for the conference, which considerably expands upon our choice of materials below.

Two other public interest groups are particularly concerned with legal and technical aspects of acid rain. At the Canadian Environmental Law Association, 5th Floor South, 8 York Street, Toronto, Ontario M5J 1R2, telephone (416) 366-9717, talk to Joe Castrilli; and at Pollution Probe, 43 Queen's Park Crescent, University of Toronto, Toronto, Ontario, telephone (416) 978-7152, talk to Bill Glenn.

Major Canadian companies involved are (a) Inco Ltd., 1 First Canadian Place, Toronto, Ontario M5X 1C4, telephone (416) 361-7511. Talk to Dr. Stuart Warner, Vice-President;

(b) Ontario Hydro, 700 University Avenue, Toronto, Ontario M5G 1X6, telephone (416) 592-3331. Ask for the Public Reference Centre;

(c) Noranda Mines Ltd., P.O. Box 45, Commerce Court West, Toronto, Ontario M5L 1B6, telephone (416) 867-7111.

Other companies with significant smelting operations in Canada include Hudson Bay Mining and Smelting Company, Ltd., P.O. Box 28, Toronto Dominion Centre, Toronto, Ontario M5K 1B8, and

Falconbridge Group, P.O. Box 40, Commerce Court West, Toronto, Ontario M5L 1B4.

## Norway

All inquiries should be directed to the office of the Norwegian Joint Research Project, begun in 1972, on the effects of acid precipitation on forests and fish. The address is: SNSF—Project, Box 61, 1432 Aas-NLH, Norway. The Research Director is Lars N. Overrein.

# Reading Materials

The following have been chosen as being representative of materials currently available. Most have bibliographies. See references in the notes as well.

Babich, Harvey et al., "Acid Precipitation: Causes and Consequences," *Environment*, Vol. 22, No. 4, May, 1980: 6-13.

Berry, Michael A. and John D. Bachman, "Developing Regulatory Programs for the Control of Acid Precipitation," *Water, Air, and Soil Pollution 8*, 1977: 95-103.

Chung, Y. S., "The Distribution of Atmospheric Sulphates in Canada and its Relationship to Long-Range Transport of Air Pollutants," *Atmospheric Environment*, Vol. 12, Nos. 6/7 and 12, 1978: 1471-1479 and 2519-2522.

Cogbill, C.V. and G.E. Likens, "Acid Precipitation in the Northeastern United States," *Water Resources Research*, Vol. 10, No. 6, 1974.

Dillon, P.J. et al., "Acidic Precipitation in South-Central Ontario: Recent Observations," *Journal of the Fisheries Research Board of Canada*, Vol. 35, No. 6, 1978.

Environment Canada, *Proceedings of a Workshop on Long-Range Transport of Air Pollution and its Impacts on the Atlantic Region*, Dartmouth, Nova Scotia, October 17-18, 1979.

Glass, Gary E. and O.L. Loucks, eds., *Impacts of Air Pollutants on Wilderness Areas of Northern Minnesota*, EPA Environmental Research Laboratory, Duluth, Minnesota, March, 1979.

Glass, N.R. et al., *Environmental Effects of Acid Precipitation*. A presentation to the Fourth National Conference on the Interagency Energy/Environmental Research and Development Program, Washington, D.C., June 8, 1979.

Harvey, H.H., "The Acid Deposition Problem and Emerging Research Needs," *Proceedings of the Fifth Annual Aquatic Toxicity Workshop*, Hamilton, Ontario, November 7-9, 1978. Fish. Mar. Serv. Tech. Rep. 862: 115-128.

Inco Limited, *Remarks at the Ontario Ministry of the Environment Public Meeting to Discuss Proposed Control Order*, Sudbury, June 4, 1980.

Jonsson, Bengt and Rolf Sundberg, "Has the Acidification by Atmospheric Pollution Caused a Growth Reduction in Swedish Forests?," *Research Notes—Department of Forest Yield Research Nr. 20*, Royal College of Forestry, Stockholm, 1972.

Kramer, J., *Susceptible Lands in Canada and U.S.A.*. A presentation to ASAP, Toronto, November 2, 1979.

Ontario Hydro, *The Sources of Acid Precipitation and Its Effects on Biological Systems*, Design and Development Division, Report No. 79164, July, 1979.

Ontario Hydro, *Atikokan Generating Station Study of Sulphate Loadings in Boundary Waters Canal Area and Quetico Park*, report prepared by Acres Consulting Services, Niagara Falls, Ontario, March, 1978.

Ontario Ministry of the Environment, *Acidic Lakes in Ontario: Characterization, Extent, and Responses to Base and Nutrient Additions*, September, 1977.

Ontario Ministry of the Environment. *Acidic Precipitation in South-Central Ontario: Analysis of Source Regions Using Air Parcel Trajectories*. May 1980.

Ontario Ministry of the Environment, *An Analysis of the Sudbury Environmental Study: Network Precipitation Acidity Data June-September, 1978*, February, 1979.

Ontario Ministry of the Environment. *Bulk Deposition in the Sudbury and Muskoka-Haliburton Areas of Ontario During the Shutdown of Inco Ltd. in Sudbury*. May, 1980.

Ontario Ministry of the Environment, *The Depression of pH in Lakes and Streams in Central Ontario During Snowmelt*, February, 1979 (revised).

Ontario Ministry of the Environment, *Notes on Acidic Precipitation for the Standing Committee on Resources Development*, February, 1979.

Ontario Ministry of the Environment, *Reclamation of Acidified Lakes Near Sudbury, Ontario*, June, 1975.

Ontario Ministry of the Environment, *Reclamation of Acidified Lakes Near Sudbury, Ontario by Neutralization and Fertilization*, undated.

Ontario Ministry of the Environment, *Sudbury Environmental Study: An Analysis of the Impact of Inco Emissions on Precipitation Quality in the Sudbury Area*, Report Number ARB-TDA 35-80, May, 1980.

Ontario Ministry of the Environment, *Survival of Rainbow Trout, Salmo Gairdneri, in Submerged Enclosures in Lakes Treated With Neutralizing Agents Near Sudbury, Ontario*, Technical Report LTS 79-2, March, 1979.

Roy, Jean-A., *Les Pluies Acides sur l'Est de l'Amérique du Nord et leurs Incidences au Québec*, Ministère de l'Environnement, Québec, undated report.

Schindler, D.W. et al., "Experimental Acidification of Lake 223, Experimental Lakes Area: Background Data and the First Three Years of Acidification," *Canadian Journal of Fisheries and Aquatic Science*, Vol. 37, 1980.

Shaw, R.W., "Acid Precipitation in Atlantic Canada," *Environmental Science Technology*, 1979: 406.

U.S. EPA, *Environmental Effects of Increased Coal Utilization: Ecological Effects of Gaseous Emissions from Coal Combustion*, Corvallis Environmental Research Laboratory, Report EPA-600/7-78-108, June, 1978.

U.S. Senate Committee on Energy and Natural Resources, *Sulphur Oxides Control Technology in Japan*, U.S. EPA Interagency Task Force Report, June 30, 1978.

Watt, W.W., "Acidification and Other Chemical Changes in Halifax County Lakes After 21 Years," *Limnology and Oceanography*, 24 (6), 1979: 1154-1161.

# Index

ROSS HOWARD is the Environmental Editor of the *Toronto Star* and has published widely on political and scientific issues. As a journalist and editor he has worked for the *Toronto Telegram*, York University, and the Ontario Ministry of Housing. He is the author of *Poisons in Public*.

MICHAEL PERLEY is the Executive Director of the Canadian Environmental Law Research Foundation and has collaborated on the research for *Poisons in Public* and Warner Troyer's *No Safe Place*.

This book accurately documents the most severe environmental problem yet in North America. It explains where acid rain comes from, what it does to plants, fish and ecosystems, how much it costs, and what the political response has been to date. In short, we're in trouble.

Part of the problem has been a lack of effective action by government and industry. But the other part is that we, the citizens of this continent, have failed to speak up or get sufficiently angry to insist that the job be done. Read this book, and you'll find it very difficult to keep quiet.

*Monte Hummel, Executive Director, World Wildlife Fund (Canada)*

Ross Howard and Michael Perley are to be commended for their fine work in putting together this important book. The documentation and analysis that it contains will aid all of us in coming to grips with this monumental problem.

*Congressman John D. LaFalce (New York), Member, Committee on Banking, Finance and Urban Affairs*

This is a fine piece of work—authoritative, imaginative, well-researched. Even lucid. It's the first book on acid rain which feels important and complete.

*Stephen Lewis, former leader of the Ontario New Democratic Party*

Citizens in both the United States and Canada should be grateful for this well-written, informative, and somewhat frightening look at one of North America's most serious environmental issues.

*Gregory S. Wetstone, Environmental Law Institute, Washington, D.C.*